Ferromagnetodynamics

Ferromagnetodynamics

The dynamics of magnetic bubbles,
domains and domain walls

T. H. O'Dell

Reader in Electronics,
Department of Electrical Engineering,
Imperial College of Science and Technology,
University of London

A HALSTED PRESS BOOK

JOHN WILEY & SONS
New York

First published in Great Britain 1981 by
The Macmillan Press Ltd

538.44
O23f

Published in the U.S.A. by
Halsted Press, a Division of
John Wiley & Sons, Inc.
New York

Printed in Hong Kong

Library of Congress Cataloging in Publication Data

O'Dell, Thomas Henry.
  Ferromagnetodynamics : the dynamics of magnetic
bubbles, domains, and domain walls.

  "A Halsted Press Book."
  Bibliography: p.
  Includes indexes.
  1. Magnetic bubbles. 2. Domain structure.
3. Ferromagnetism. I. Title.
QC754.2.M34032 1981   538'.44   80-25331
ISBN 0-470-27084-5

Contents

Preface

This book was started during a most interesting visit to the Central Research Institute for Physics, Hungarian Academy of Sciences, Budapest, early in 1977. There, the author was introduced to the powerful experimental technique of high-speed microphotography, which has played such an important part in the development of magnetic bubble domain dynamics, by Dr G. J. Zimmer and his colleagues, Dr L. Gál and Dr G. Kádár. The financial support of the Institute and Imperial College during this visit is gratefully acknowledged.

In completing this book, the author has drawn on discussions and correspondence with many people and this should be reflected in the bibliography which, it is hoped, is as complete as possible. Particular thanks are due to Professor E. P. Wohlfarth, who has always been a source of good advice and encouragement, and to Dr K. D. Leaver, who was very helpful on the ripple problem.

For line drawings which have a citation in the caption, permission to reproduce or adapt has kindly been given by the authors cited and by the copyright holders: The American Institute of Physics, The Bell System Technical Journal, The Institute of Electronic and Electrical Engineers, The Institute of Physics, North Holland Publishing Co., Amsterdam, and Springer-Verlag, Heidelberg.

Finally, sincere thanks to Mrs Joan Jeffery for her skilful preparation of the typescript.

T. H. O'DELL

1 The History and Experimental Techniques of Ferromagnetodynamics

1.1 Introduction

Ferromagnetodynamics is the study of the way in which the magnetisation of a ferromagnet can be changed in both space and time. Experimentally, the subject really began just after 1930, when the idea of ferromagnetic domains was very new and very little was known about the way in which the magnetisation could change with time. After the Second World War, our knowledge of magnetism had advanced very considerably and experimental work became far more exact when single crystals of magnetic materials, both metals and insulators, became available. It was then possible to make very definite assumptions about the spatial distribution of the magnetisation and concentrate on a measurement of its time dependence under the action of a time dependent applied field.

Around 1950, it became possible to discuss the experimental work from the point of view of existing theory. At the same time, the closely related fields of magnetic resonance and spin waves were undergoing rapid development. Magnetic resonance and spin waves involve changes in the magnetisation which are relatively small compared to the magnetisation itself and the time scales involved are usually shorter than 10^{-9} s. Ferromagnetodynamics, on the other hand, is concerned with very large changes in the magnetisation, a complete reversal for example, occurring over distances which may be well under 10^{-6} m but the time scales involved are usually longer than 10^{-9} s.

In this chapter we shall review the early experimental work briefly and then outline the development of the subject from the experimentalists' point of view. This will help to fix the order of magnitude of the scales we are involved with, in both space and time, and the kind of experimental facts which need explanation. It will be seen that the subject of ferromagnetodynamics has always been under pressure from technology. The application of magnetic materials in electronics, for microwave devices, for computer memories and, for a very long time, in the realisation of high frequency inductors, has always kept magnetism an active and developing field. In heavy electrical engineering we find the same importance attached to magnetic materials and this has had a great influence on the development of new alloys.

In the following chapters, we shall try to describe and develop the theory of ferromagnetodynamics and apply it to the experimental work. The theory itself has only recently become advanced enough for us to understand some of the most simple experimental results and the reason for this advance is certainly the magnetic bubble domain. The developments in bubble domain technology have given us, for ten years now, single crystal materials in which we can actually see the changes in magnetisation with a resolution of better than 10^{-6} m and 10^{-8} s in space and time. The experimental results coming from the large number of workers in the field of magnetic bubbles, all over the world, has given a considerable impetus to theoreticians and, although the theory has advanced rapidly, the story is by no means over. It is hoped that this book will give an outline of the established theory and serve as a guide to the almost overwhelming literature which now exists on the subject of ferromagnetodynamics.

1.2 Early Experimental Work

The title of a long paper by Lyle and Baldwin (1906) 'Experiments on the propagation of longitudinal waves of magnetic flux along iron wires and rods' tells us a great deal about the attitude of the late 19th and early 20th centuries to ferromagnetism. In those days, ferromagnets were considered to be homogeneous media which could be characterised by two constant parameters; permeability, μ, and conductivity, σ. From the experimentalists' point of view, it was interesting to see if changes in the magnetisation in one part of a magnetic body would diffuse to another part, according to the well established Maxwell theory which tells us that the 'magnetic flux density', B, as it was then called, would be governed by the diffusion equation,

$$\nabla^2 B = \mu\sigma \ \partial B/\partial t \qquad (1.1)$$

Lyle and Baldwin (1906) made experiments in which they excited the centre of a long iron wire by means of a coil wound around it which carried an alternating current. They then explored what was happening along the length of the wire with a search coil and came to the conclusion that things were a great deal more complicated than equation 1.1 would suggest. They gave five references to earlier work, the earliest being Oberbeck (1884), who made very similar experiments, and an interesting reference is Zenneck (1902) whose experiments were concerned with toroidal samples of iron which had exciting coils wound over a very small section of their circumference.

The concept of magnetic domains, which was introduced by Pierre Weiss in his famous paper on the molecular field theory of ferromagnetism (Weiss, 1907) was not taken up by experimentalists until Barkhausen published his work in 1919. This work showed that the magnetisation could change in a very discontinuous way, giving rise to the well known 'Barkhausen effect'. Two papers by van der Pol (1920), which can be found among his selected scientific papers (Bremmer and Bouwkamp, 1960) give considerable insight into the way in which the domain concept began to be introduced into ferromagnetism. The model was one in which the demagnetised state was a disordered array of very small regions, the domains, within the material and magnetisation due to an applied field involved these domains forming thread like chains, each domain in the thread having its magnetisation flipped around to point along the direction of the thread, which was the direction of the applied field.

Very strong experimental evidence for this model was given by de Waard (1927), who made calculations of the way in which the magnetisation of such ensembles of domains would vary with the applied field and then compared his models with experiment. The most surprising development of all was in 1931 when Francis Bitter published the results of his first 'Bitter pattern' observations (Erber and Fowler, 1969). The patterns were interpreted as further evidence for thread like domains.

This situation seems to have persisted until 1935 when the famous paper by Landau and Lifshitz (1935) was published. This paper is of such importance to our subject that chapter 2 is concerned almost entirely with it. A glance at the first paragraph of this paper confirms the authority behind the thread domain idea. A further review of the development of domain theory a few years later by Kennard (1939) gives many further references.

Landau and Lifshitz (1935) introduced the idea that the magnetisation could change by a movement of the boundary between domains, that domains magnetised in the direction of the applied field would expand at the expense of domains magnetised against the applied field. We find this idea being taken up rapidly, first at a meeting in Göttingen in 1937 (Becker, 1938) and then by Kondorsky (1938) and by Brown (1939). However, as so often happens, and this is certainly the case in ferromagnetodynamics today, the experimentalists were leading the theoreticians because it was not Landau and Lifshitz who were the common reference of the last three cited authors but Sixtus and Tonks and their experiments. These experiments mark the real beginning of ferromagnetodynamics and are described in the next section.

1.3 The Experiments of Sixtus and Tonks

These experiments are described in a series of five papers: Sixtus and Tonks (1931, 1932), Tonks and Sixtus (1933a, 1933b) and Sixtus (1935). Some further work was reported by Sixtus at the 1937 meeting in Göttingen (Becker, 1938), which was referred to in the previous section.

The apparatus is shown in figure 1.1. This was centred around a nickel-iron alloy wire. The alloy was chosen so that when the wire was put in tension, the magnetostriction acted with the shape anisotropy of the wire to make the remnant state one single ferromagnetic domain with its magnetisation directed along the axis of the wire. This remnant state was achieved in the experiment by applying a large bias field, using the bias field coil shown. The bias field was then reduced, through zero, to be applied in the reverse direction but only increased to a small value so that it was below the value required to reverse the magnetisation of the wire by the spontaneous nucleation of a reverse domain.

Magnetic reversal of the wire was then achieved in a controlled way by passing a current through the nucleating field coil, shown in figure 1.1, so that the reverse bias field was increased in magnitude in that region of the wire. In later experiments this nucleation of a reverse, or seed, domain was done in a more definite manner by using a pulse of current through the nucleating field coil. The nucleated reverse domain then expanded, under the influence of the small applied bias field, and the propagation of one end of this expanding domain along the wire could be observed by means of the voltage pulses which were induced in the two pick-up coils shown in figure 1.1. By measuring the time difference between the two pulses, for various spacings and positions of the two coils, it was possible to work out the velocity of propagation of the domain boundary. The velocity was found to depend upon the applied bias field and

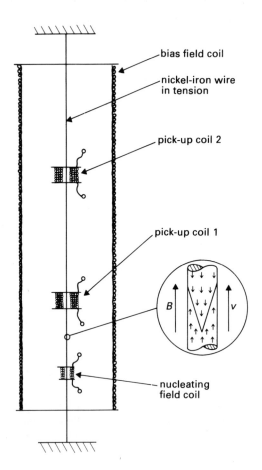

Figure 1.1 The apparatus used by Sixtus and Tonks (1931). The magnetisation of a nickel–iron wire reverses by the propagation of a domain boundary of the kind shown inset

the stress in the wire. It did not vary along the length of the wire. Some typical results are shown in figure 1.2.

The striking point about the results shown in figure 1.2 is the very linear relationship between the velocity and the applied field. This does, however, only apply over a very small range of applied field because quite a large field, B_0, must be applied before any propagation can be observed. The maximum field that can be used is the one which would cause spontaneous reversal and this was found to be not very much greater than B_0. This linear relationship was expected by Sixtus and Tonks because they considered that the only resistance to the motion of the magnetic discontinuity would come from the coercivity of the material, which would explain B_0, and then from the eddy-current loss induced in the wire by the moving magnetic discontinuity which would give rise to a viscous damping term, linear in velocity.

In order to compare their experimental results with theory, Sixtus and Tonks needed some model of the moving domain wall. This was inferred from the waveform of the voltage pulse induced in the pick-up coils as the domain wall passed through and the conclusion was that the boundary was a cone-like one, as shown inset in figure 1.1. We thus see that Sixtus and Tonks were proposing that a change in the magnetisation would take place by domain wall motion some years before the domain wall model had developed. In figure 1.1, the angle of the

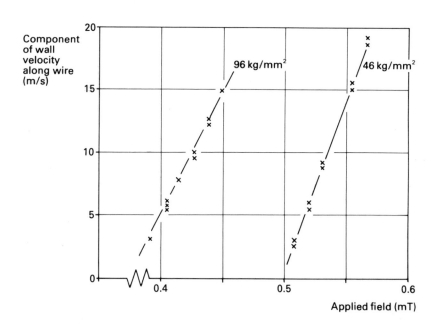

Figure 1.2 Typical results copied from Sixtus and Tonks (1931) for a 14/86, Ni/Fe, wire, 380 μm in diameter, under tension. The tensile stress is shown

cone-like boundary has been exaggerated for clarity. The cone does, in fact, occupy a length of wire which is many times its diameter.

More detail may be seen if we consider a recent application of the technique developed by Sixtus and Tonks. O'Handley (1975) has used the technique to study domain wall kinetics in wire samples of the ferromagnetic glass $Fe_{76}P_{12}C_7Cr_{4.5}B_{0.5}$ and his results are shown in figure 1.3. By observing the waveform of the voltage pulse induced in his pick-up coils, O'Handley was able to see that the length of the cone discontinuity was almost constant, independent of velocity and position, at 25 mm, over 200 times the diameter of the wire. Figure 1.3 shows the same general features as figure 1.2. There is a very small positive intercept, B_0, when we extrapolate the linear velocity–applied field points back. The results can be expressed by the relationship

$$v_n = \mu_w \ (B - B_0) \tag{1.2}$$

where v_n is the velocity of the conical domain wall normal to its surface, B is the applied field, B_0 has been discussed above and μ_w is what we would now

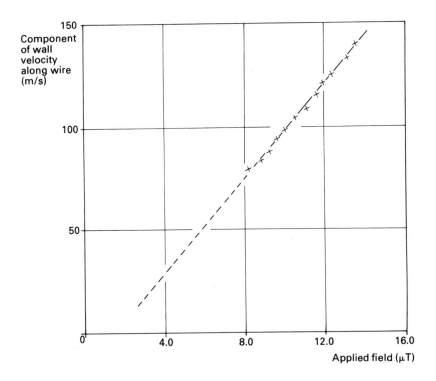

Figure 1.3 Results copied from O'Handley (1975) for a wire of the ferromagnetic glass $Fe_{76}P_{12}C_7Cr_{4.5}B_{0.5}$, 115 μm in diameter

call a domain wall mobility. Equation 1.2 is going to come up again and again throughout this book because so many experimental results may be expressed in this way.

The normal velocity of the wall, v_n, is, of course, much smaller than the velocity observed in the Sixtus and Tonks experiment because of the very small cone angle of the moving magnetic discontinuity. The value for μ_w in the case of O'Handley's experiments, shown in figure 1.3 is 2.7×10^4 (m/s) per T and he was able to obtain reasonable agreement with the theory of wall motion for insulating magnetic materials, which is dealt with here in chapters 2 and 3, combined with the theory for wall motion in conducting media, involving simple eddy-current damping, which is considered in chapter 5. Sixtus and Tonks were not able to get very good agreement with their own ideas for an eddy-current damping model but this was mainly because, while they had made the quite new and correct proposal that the magnetisation was changing by wall motion, they had not got any clear model for the structure of the wall itself. It is to this point concerning wall structure which we should return for the continuation of an historical account of ferromagnetodynamics.

1.4 The First Dynamic Experiments with Single Crystals

The paper by Landau and Lifshitz (1935), which was referred to in section 1.2, had not only introduced the idea of domain wall motion but had dealt with the atomic scale structure of a particularly simple domain wall. This is shown in figure 1.4 and represents the solution obtained by Landau and Lifshitz for the wall between two domains in a material which has only one preferred direction of magnetisation, the z-axis in figure 1.4. Landau and Lifshitz showed that one possible solution to the problem of the wall structure is that the atomic magnetic moments should rotate about the normal to the plane of the wall and that they should always lie in the plane of the wall as they rotate. The magnitude of the elementary magnetic moments is the same everywhere, only the direction changes. The distance over which this rotation takes place depends, of course, upon the material being considered but in the majority of materials the rotation is more or less completed over about one hundred atomic spacings. A full treatment of the Landau–Lifshitz wall, and some discussion of why it is so often referred to as the Bloch wall, will be given in chapter 2.

Once this wall structure had been described by Landau and Lifshitz (1935) it was naturally considered by experimentalists and attempts were made to set up the conditions where an isolated wall of this particularly simple kind could be subjected to an applied field, along the z-direction in figure 1.4, and its resulting motion examined. Landau and Lifshitz had given a model for the resistance to wall motion due to the fundamental spin relaxation phenomena they expected, which is the main topic of chapter 2 here, and the eddy-current damping effects could be calculated accurately once the wall structure was assumed.

The moving wall in the Sixtus and Tonks experiment would approximate to

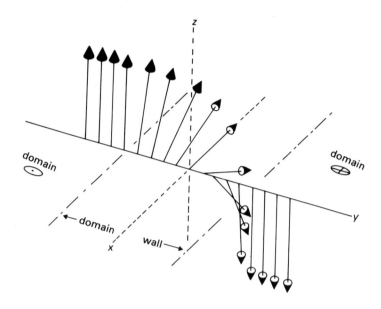

Figure 1.4 The domain wall proposed by Landau and Lifshitz (1935) between two domains in an infinite material having an easy direction along z. The elementary or atomic magnetic moments are of constant magnitude and rotate in (x, z) as we go along y from one domain to the other

the structure shown in figure 1.4 if the cone angle shown inset in figure 1.1 was very small and if the axis of the wire was a strongly preferred direction of magnetisation. With a polycrystalline wire this latter point is difficult to achieve and it was natural that experimentalists should consider the possibility of using single crystal materials. These were known to have well defined easy directions of magnetisation along particular crystallographic directions.

The first successful dynamic experiments of this kind were by Williams, *et al* (1950) who used single crystals of $Si_{0.04}Fe_{0.96}$ and obtained excellent agreement between their experimental results, which again followed the relationship given by equation 1.2 and their calculation of wall mobility for eddy-current damping in this conducting ferromagnetic material. The later experiments of Galt (1952) with single crystals of Fe_3O_4 and then with single crystals of the ferrite $Ni_{0.75}Fe_{2.25}O_4$ (Galt, 1954) are more interesting and involve the same experimental techniques, so we shall describe Galt's work with ferrite single crystals.

There had been some very significant developments in magnetism during the nineteen years which separate the publication of the paper by Landau and Lifshitz (1935) and Galt's experiments. One of the most important of these was the development of the ferrite materials, the first full description of this work being by Snoek (1947). Galt (1954) chose one of these new materials for his later

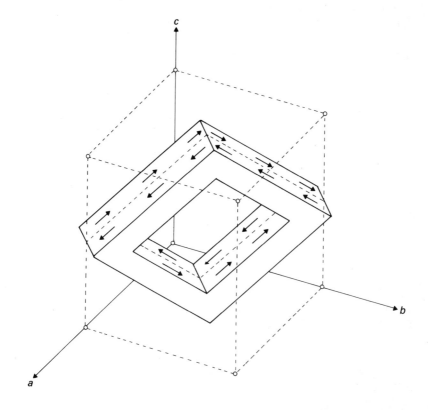

Figure 1.5 The way in which the 'picture frame' sample used by Galt (1954)
was cut from a crystal of cubic $Ni_{0.75}Fe_{2.25}O_4$ (Copyright 1954,
American Telephone and Telegraph company. Reprinted by permission.)

experiments. He had previously been involved in the successful growth of single
crystals of ferrites (Galt *et al.,* 1950) and the kind of single crystal sample he
used is shown in figure 1.5.

The ferrite, $Ni_{0.75}Fe_{2.25}O_4$, has a cubic crystal symmetry and the easy directions of magnetisation lie along the (111) directions. By cutting the 'picture
frame' sample from a single crystal, shown in figure 1.5 orientated to the cubic
axes (*a, b, c*), a closed magnetic circuit was obtained. The lowest energy
demagnetised state of such a sample should then be one involving only a single
domain wall parallel to the (111) plane which has the smallest total area. Of the
two possibilities, the one shown in figure 1.5 can be made the preferred one by
making the thickness of the sample greater than the width of the frame.

To obtain this single wall state requires a very perfect single crystal because
the difference in the total energy of a sample when it contains a single wall
compared to when it contains two or three is very small. Once a single wall

state is achieved, however, it is fairly stable because the nucleation of a second wall does call for quite a large energy input. The single wall state could be identified by Galt (1954) in the same way as it had been done in the earlier work by Williams *et al.* (1950). This was to use the decoration technique discovered by Bitter, which was discussed in section 1.2 and involves a thin layer of magnetic colloid on the surface of the sample to show up the position of the domain wall by interacting with the stray field of the wall. Galt prepared a number of samples and examined these with the colloid technique to select one which showed the single domain wall running entirely around the sample on the outside faces.

The sample was then cleaned and wound with a uniform drive coil so that a magnetic field could be applied along the preferred direction of magnetisation by supplying this coil with a constant current pulse. The motion of the single domain wall was inferred from the voltage induced in pick-up coils wound around different parts of the sample. Knowing the saturation magnetisation and the sample dimensions it was possible to work out the velocity of the wall, assuming that only a single wall was still present under dynamic conditions. A very good check was possible on this point, however, because the induced voltage was observed to double when the drive field exceeded a certain critical value, showing that a second wall had been nucleated.

Figure 1.6 shows some of the results given by Galt (1954) and we see that these again have the form of equation 1.2. In contrast to our previous figures

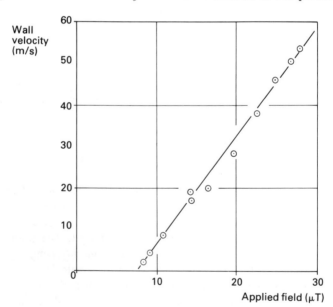

Figure 1.6 The mobility data of Galt (1954) taken on a picture frame sample of $Ni_{0.75}Fe_{2.25}O_4$ at 201 K (Copyright 1954, American Telephone and Telegraph company. Reprinted by permission.)

1.2 and 1.3, the velocity in figure 1.6 is the true velocity of the wall, v_n, so that μ_w, in equation 1.2, is simply the slope of the line. Galt found that the wall mobility, μ_w, was 2.6×10^6 (m/s) per T and was able to show that this agreed well with the existing theory, which will be discussed in detail in chapter 2. The value of B_0 he obtained also seemed very sensible because it compared well with the normal coercive force of the sample. A further encouraging experimental fact was that very low velocities could be observed at applied fields just in excess of B_0. This is certainly not the case for the results shown in figure 1.3, for example.

The general conclusion from Galt's work, at that time, was that this particularly simple kind of wall motion was now well understood. The experiments were soon repeated on a different ferrite, $Mn_{1.4}Fe_{1.6}O_4$, by Dillon and Earl (1959) and good agreement between theory and experiment was also obtained with this material.

Shortly after Neilson and Dearborn (1958) solved the problem of growing single crystals of the garnet $Y_3Fe_5O_{12}$ (YIG) the experiments could be repeated using picture frame samples of this very low loss and almost perfectly insulating magnetic material. This was done by Hagedorn and Gyorgy (1961), Wanas (1967) and with great accuracy by Harper and Teale (1969). The experimental results of Harper and Teale are shown here in figure 1.7 and a glance at these, in comparison with figure 1.6, shows that the wall velocity in YIG appears to be much

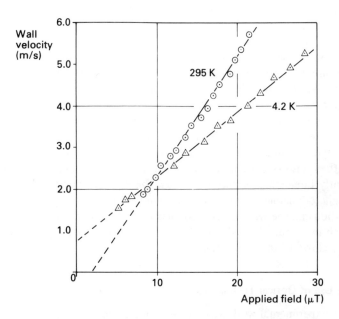

Figure 1.7 Data copied from Harper and Teale (1969) taken on a picture frame sample of $Y_3Fe_5O_{12}$ at the two temperatures shown

lower, for the same order of applied field, than it is in nickel ferrite. This was a most unexpected result, because the theory at that time showed that the mobility of a material should increase as the magnetic losses were reduced, and single crystal YIG is the lowest loss material we have, even today. Other problems can be seen in figure 1.7. We do not observe any low velocities in YIG. At 295 K we can go down to just below 2 m/s but the drive field is still five times the inter-cept value, B_0. At 4.2 K something is clearly wrong with extrapolating the results to low fields along a straight line, because this would give a negative value of B_0. There must be a set of experimental results in between some very small value of applied field and the beginning of the straight lines shown in figure 1.7 but, for some reason, we cannot observe this. These problems will be dealt with in chapter 2.

To conclude this section on the single crystal experiments, of the kind which use inductive pick-up coils to infer what the magnetisation is doing, we must mention the very interesting work of Tsang *et al.* (1975, 1978) on single crystals of $YFeO_3$. This material is an orthoferrite and is a weak ferromagnet, which means that its magnetisation is due to two very strongly coupled magnetic sub-lattices which are equal in magnitude and almost antiparallel. In $YFeO_3$ the net magnetisation has a very strongly preferred direction along the c-axis.

Tsang *et al.* (1975, 1978) used the method of Sixtus and Tonks, described in section 1.3, with long single crystal bars which had the a, b or c-axes along their length. For the c-axis case, the Sixtus and Tonks apparatus shown in figure 1.1 can be used but for the other two orientations the bias field, and the orientation of all the other coils, must be rotated through 90°.

The results given by Tsang *et al.* (1975, 1978) have been combined in figure 1.8. Again we see results which may be expressed as a linear relationship, equation 1.2, but only the c-axis results seem to fit a simple model in which the intercept, B_0, is positive and low velocities may be observed just above B_0. The results for the a and b-axes have a negative intercept and this must mean that some dramatic change takes place in the wall structure before it can begin to move. The wall structure itself is complicated in this material, because it is a weak ferromagnet, and a different wall structure is expected for the a and b cases anyway. We shall discuss these results in more detail in chapter 2. The results of Tsang *et al.* are particularly interesting because they show the highest velocities yet observed in ferromagnetodynamics. We have only shown the linear region of the results here. Above 4000 m/s the wall velocity couples with the sound velocity in the crystal and much more complicated behaviour is seen. Tsang *et al.* were able to observe velocities above 12 000 m/s in these experiments.

1.5 The Use of Optical Techniques

In all the experimental work we have described so far the changes in the magnet-isation have been determined inductively, that is, by looking at the changes in the magnetic field produced by the magnetisation, using some kind of pick-up

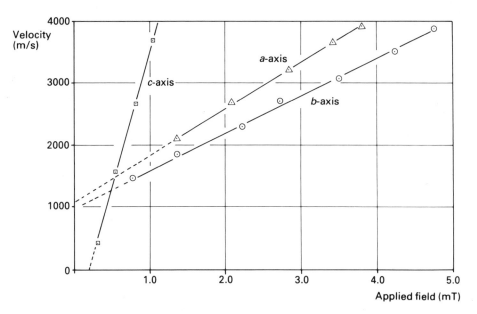

Figure 1.8 Data copied from Tsang and White (1974) and from Tsang *et al.* (1975, 1978) using the Sixtus and Tonks technique on single crystal bars of $YFeO_3$. The results are labelled according to the crystal axis which lay along the length of the bar

coil, and then working out what the magnetisation was doing by assuming some particularly simple domain structure.

These techniques were clearly open to some ambiguity and there was a very good reason to try some more direct method. The need for a more direct technique became rather urgent after 1956 because there had been a very import-ant technical development in magnetism at that time. This was the beginning of attempts to make very high-speed memories for digital computers out of thin magnetic films, an interesting story in itself which we shall discuss in chapter 5 when we look at the ferromagnetodynamics of conducting materials.

The suggestion that thin magnetic films, of alloys like NiFe prepared by evaporation or electrodeposition, might show 'considerable promise as core materials for magnetic memory matrices in high speed digital computers' was made in a paper by Conger and Essig (1956). The idea was supported by some earlier work (Conger, 1955) which had shown that very rapid switching of the direction of the magnetisation in a thin film was possible and this was thought to be due to the fact that domain walls would not form in very thin films, say less than $0.2\,\mu m$ thick, a point which had been made earlier by Kittel (1946b). A theoretical paper by Kikuchi (1956) which claimed to show that it should be possible to switch the direction of magnetisation in thin films in times as short as 2×10^{-9} s also excited a great deal of interest.

The computer memory of that time was the magnetic core store originally proposed by Forrester (1951). This was used as the main high-speed random access memory of the machine and had a cycle time, the time needed to read a word from the store and put it back again, of a few microseconds. An order of magnitude improvement in this cycle time, if the physical size of the memory could be kept the same, or reduced, would have been a very important step forward in computer technology and it was this which the thin film memory seemed to promise.

That very thin evaporated films of nickel–iron did switch in times as short as a few nanoseconds was first shown experimentally by Smith (1958). Other workers had difficulty in reproducing Smith's results but this was done without question by Dietrich *et al.* (1960). Thin film memory development then took on almost epidemic proportions and ended up with quite a reasonable system in the various kinds of plated wire memory. An excellent review of this work is given by Mathias and Fedde (1969) who give a large bibliography. The flat kind of film memory faded into obscurity because it could never achieve a bit density compatible with its potential speed. In any case, the field of random access computer memories was being rapidly taken over by semiconductors.

The work with thin films gave experimentalists a strong incentive to develop some technique which would show them directly what the magnetisation was really doing when it changed direction in the film. The normal state of magnetisation in these films was that they were a single domain, magnetised in the plane of the film along some easy direction. Humphrey (1958) had been able to show, by using pick-up coil techniques, that very fast reversal of this magnetisation took place by some kind of complete rotation of the magnetisation vector in the plane of the film, a point we shall look at theoretically in chapter 5, while the slower reversals took place by domain wall nucleation. An obvious suggestion for an experimental method which would show what was happening was to use one of the magneto-optic effects and this was first done dynamically by Lee and Callaby (1958) using the Kerr effect. Two small rectangular spots of polarised light were produced on the film and the reflected light from these was monitored with a photomultiplier. The passage of a single domain wall could be observed as it passed, first through one spot and then the other, the velocity of the wall being determined from the time difference between the two events and the distance between the two light spots. This technique was used by Ford (1960), who used a single spot of light, of known diameter, and measured the time taken by a moving wall to cross the spot. He compared measurements of the wall velocity inferred from photographs of the wall position taken before and after a short drive pulse. It is very interesting to find Ford (1960) saying that this latter method, compared to the former, gave: 'erratic results, probably because the walls do not begin moving with the start of the pulse and move for some unknown distance after the end of the pulse'. We shall find this idea coming up again in very recent times when we look at magnetic bubble dynamics.

This kind of optical technique for ferromagnetodynamics, in which a small spot on the sample is illuminated and the changes in magnetisation at that spot are monitored by means of the Kerr effect, was brought to a very high state of perfection by Thompson and Chang (1966). These workers were able to reduce the spot size to 25 μm diameter and the time resolution was a few ns. Switching in thin films of permalloy could be studied with this apparatus, the position of the spot showing different behaviour at different points on the film. The magnetic switching process had to be repetitive, however, because the Kerr effect signal had to be recovered from the system noise by sampling techniques. It is important not to confuse the techniques described here, where the intention is to obtain high spatial resolution, with the much simpler optical techniques applied to bubble domain materials using the Faraday effect (Seitchik *et al.,* 1971, O'Dell, 1973 and Argyle and Halperin, 1973). These methods use an optical technique in the same way as the earlier experiments used search coils; to measure the time variation of the magnetisation averaged over quite a large area.

1.6 The First Use of High-speed Photography in Ferromagnetodynamics

The success of the optical techniques, described in the previous section, obviously suggested the development towards some kind of optical technique which would show the detail in the time varying magnetisation over an extended area of the thin metal films being studied at that time.

Good quality photographs of the magnetic domain structure in thin films had been made by Green and Prutton (1962) and Copeland and Humphrey (1963) had been able to take a series of photographs showing the way in which very slow switching took place in a thin permalloy film. The drive pulse could be interrupted during very slow switching and the intermediate domain patterns, which are stable in very slow magnetic switching, could be photographed with exposure times of several seconds. These long exposure times are needed, with normal kinds of microscope light source, because the contrast between domains of different directions of magnetisation is only due to the very small difference in the angle of polarisation of the reflected light. This means that the analyser at the output of the microscope must be set near to extinction for domains of one polarity and the light coming from the other domains is very weak indeed. In order to use exposure times of only a few nanoseconds, the time scale of ferromagnetodynamics, we must either increase the incident illumination very considerably or use some kind of image intensifier between the output of the microscope and the camera. Modern systems use both solutions.

The first high-speed photographs of ferromagnetodynamics were taken by Conger and Moore (1963) of the magnetic reversal process in a 0.1 μm thick film of NiFe and showed this taking place by the growth of a reverse domain at the centre of the film. The quality of the pictures was quite good and were made using a colour contrast Kerr effect which involved a thin layer of ZnS deposited upon the surface of the magnetic film.

The apparatus used by Conger and Moore (1963) was remarkable. The light source was sunlight, from a sun tracking mirror, which went to a multifaceted mirror rotating at 10^6 rpm. This mirror was driven by a special helium turbine. The light pulse so obtained was 100 ns long. Single exposures were not possible, because the light intensity was too low, so that the magnetic film reversal was run repetitively and the time delay between the beginning of the drive pulse and the light flash was set at some particular value so that a long exposure photograph could be taken which would be a superposition of several thousand light pulses. By varying the delay, it was possible to obtain a series of photographs showing the magnetisation reversal process, provided this process was truly repetitive. In other words, this was a stroboscopic technique, not really high-speed photography. The same kind of stroboscopic technique was used later on orthoferrites, with the Faraday effect, by Rossol (1969), who used a helium–neon laser with an electro-optic modulator, and by Ikuta and Shimizu (1973) who used a light emitting diode.

The first really single-shot high-speed photographs of ferromagnetodynamics were taken by Kryder and Humphrey (1969a, b, 1970) who used a 10 ns pulse from a ruby laser to obtain Kerr effect micrographs of the changing magnetisation in thin films of nickel–iron. The experimental difficulties were considerable (Kryder and Humphrey, 1969b) particularly the time jitter between the laser flash and its trigger pulse which increased as the ruby aged, this having to be replaced after only 5000 flashes. The pictures obtained were of good quality, however, (Kryder and Humphrey, 1970) and showed the detail of the very complicated processes which could occur during the reversal of the magnetisation in a thin metal film. By this time, however, thin metal films were no longer of great interest because attention was turning towards the new magnetic bubble domain materials. The foundations laid by Humphrey's group, at the California Institute of Technology, for the development of high-speed photography of ferromagnetodynamics (Humphrey, 1975), were soon to be built upon for this new bubble domain work and we shall describe this after looking at magnetic bubbles in general.

1.7 Magnetic Bubble Domain Materials

In 1967 there was another very important technical development in magnetism which was to precipitate the same scale of activity that thin films had done over ten years before. This was the publication by Bobeck (1967) of his work on magnetic bubbles in the orthoferrites and their possible application in computer memories and for a purely magnetic logic. The early history of magnetic bubbles is given in the books by O'Dell (1974), Bobeck and Della Torre (1975) and Chang (1975) and this technology left the laboratories for production in 1976 and is now used in computers for mass memory, replacing small disc files and tape cassettes. Work on more advanced memories, particularly towards higher bit density and speed, continues to be a very active field, however, and the realisation of the full capabilities of bubbles in computing is only just beginning.

Before giving a brief review of the main steps in the history of bubble domain dynamics we should pause to look back at the work we have already described to bring out the really important feature which makes the ferromagnetodynamics of large single crystals, thin metal films and bubble domain materials fall into three very different categories. This feature is the structure of the domain wall, the boundary between two domains, the motion of which is the problem of real interest in the majority of ferromagnetodynamic problems.

Let us return for a moment to section 1.4, where we discussed the first dynamic experiments with single crystals. Figure 1.4 shows the structure of the domain wall proposed by Landau and Lifshitz (1935) and it is this kind of wall which the picture frame experiments were intended to involve.

This wall is shown again in figure 1.9a, this time on the scale of the sample and with just the directions of the magnetisation for the domains and for the centre of the wall shown for clarity. This is the kind of wall we would expect in a picture frame sample, figure 1.5, with the magnetisation in the domains lying in the plane of the sample. The width of the domain wall has been exaggerated considerably in figure 1.9a, for clarity. In nearly all practical cases the sample would be about 1 mm thick while the width of the wall would be less than 1 μm.

The Landau–Lifshitz solution to the wall structure was, in fact, for a magnetic medium of infinite extent. When we are dealing with the case represented by figure 1.9a, a sample 1 mm thick and walls less than 1 μm wide, the Landau–

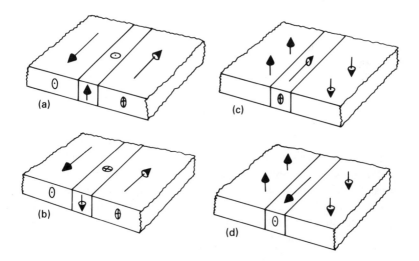

Figure 1.9 **The domain wall in a sample where the easy direction of magnet-isation lies in-plane can take an anticlockwise screw-sense (a) or a clockwise one (b). The same applies in a bubble domain material where the easy direction of magnetisation is normal to the surface, (c) is an anticlockwise screw-sense wall and (d) is clockwise**

Lifshitz solution is a fair approximation. The magnetic field due to the magnetisation itself is contained within the material everywhere, except over the very small region where the wall meets the surface of the sample. Quite a large magnetic field will be found there and it is this field which is responsible for forming the Bitter patterns which were so important in the preparation of the picture frame samples discussed in section 1.4.

The Landau–Lifshitz solution certainly does not apply when the thickness of the sample becomes small enough to be comparable with the wall width. This is the case for the thin evaporated films, which were discussed in the two previous sections. In such thin films the structure of the wall is very much more complicated than the simple Landau–Lifshitz solution and this problem will be discussed in chapter 5 when we look at the ferromagnetodynamics of such thin films.

To return again to the simple wall, it should be noted that there are two possible forms for this structure, depending upon the screw-sense. Figure 1.4 shows a wall in which the magnetisation rotates anticlockwise about the y-axis as we proceed along the y-axis. This is the situation in figure 1.9a. The opposite screw-sense is, of course, also a solution of the equations, which will be given in full in chapter 2, and this clockwise rotation is shown in figure 1.9b. We shall find that we may have conditions in which these two kinds of wall structure have different dynamic properties.

Let us now look at the magnetic bubble domain materials where we have a different situation. This is shown in figures 1.9c and d, representing the two possible screw-senses for this wall, assuming to begin with that it has the simple Landau–Lifshitz structure shown in figure 1.4. In the bubble domain materials, the preferred direction of magnetisation is normal to the plane of the sample, which is a thin layer a few microns thick. In contrast to the walls shown in figures 1.9a and b, there is now a strong magnetic field near the surface of the sample which is perpendicular to the plane of the wall. It follows that we cannot expect the Landau–Lifshitz solution to apply near the surface but it might apply well inside the sample, always assuming the thickness of the sample is very much greater than the domain wall width.

The conclusion is that the dynamics of domain walls in materials where the magnetisation lies in the plane of the sample, figures 1.9a and b, will be different to the dynamics of walls in the bubble type of material where the magnetisation is normal to the plane of the sample. For this reason we treat the two cases separately, using chapter 2 to cover the first and chapter 3 for the second. This is just for the simplest problems involving straight domain walls. For bubble domain dynamics we must turn to chapter 4. Here we shall continue to look at the history of ferromagnetodynamics by looking at the significant experimental results which were published once work with magnetic bubbles really got going.

1.8 The Bubble Collapse Experiment

An isolated magnetic bubble is sketched in figure 1.10. The vast majority of

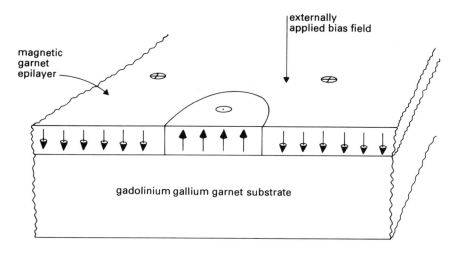

Figure 1.10 The magnetic bubble domain

experiments are done on epitaxial garnets, grown by liquid phase epitaxy on a garnet substrate, the easy direction of magnetisation being in the direction of growth. The garnet epilayer is typically a few microns thick and the bubble domain is a cylindrical domain which extends right through the epilayer and has its magnetisation opposed to an externally applied bias field.

If the bias field is increased, the bubble domain gets smaller, its circular shape being maintained by what is really equivalent to surface tension, so that the use of the name bubble is apt. This surface tension comes from the high energy density of the domain wall, due to the fact that the atomic scale magnetic moments, within the wall, must turn away from the preferred direction and because of the exchange energy which comes in when neighbouring magnetic moments are not parallel to one another. If the bias field is increased too much the bubble will collapse because the contracting force due to the surface tension will become strong enough to overcome the magnetic pressure, due to the large magnetic field inside the bubble, which is trying to make the bubble expand.

The first detailed experiments on the dynamics of isolated bubble domains were described by Bobeck *et al.* (1970) and depended entirely upon the ease with which a micron sized bubble domain could be seen in a microscope, whereas there was no hope of detecting the magnetic field it produced. This was the bubble collapse experiment in which a pulsed magnetic field was applied to an isolated bubble. The pulsed field added to the bias field and its amplitude, or duration, T, was increased until the bubble domain was observed to collapse when viewed in the microscope. If a simple wall mobility model is used, as given by equation 1.2, an almost straight line relationship between T^{-1} and the pulse amplitude is expected, once the pulse amplitude exceeds the value for static collapse by a factor of about 2. The theory of the bubble collapse experiment is given by Callen and Josephs (1971) and O'Dell (1974).

The bubble collapse experiment was taken up by a number of workers and on a wide range of materials. Bobeck *et al.* (1970) had found that their bubble collapse results for orthoferrites and for one garnet sample were consistent with a simple constant wall mobility model but their results using magneto-plumbites were very non-linear, suggesting the most complicated dynamic behaviour. The linear behaviour of the garnets was not confirmed by Calhoun *et al.* (1971) or by Callen *et al.* (1972) who both observed what appeared to be a rapid saturation of the wall velocity for a contracting bubble. The results of Vella-Coleiro *et al.* (1973, 1974) showed that the bubble collapse data was quite inconsistent with the data produced by the bubble translation experiment, which will be described in the next section, and the data from observing bubble domain motion in a device-like environment. The bubble collapse experiment was beginning to show up the difficult features of bubble dynamics which were to be familiar among experimentalists and theoreticians within a few years. The experimental difficulties became really clear when high-speed photography could be applied to bubble collapse and Humphrey (1975) published photographs showing that the collapsing bubble became much smaller than its static collapse diameter before it finally disappeared. The other fascinating results of applying high-speed photography to bubble dynamics will be described in section 1.10.

1.9 The Bubble Translation Experiment

The next major experiment in bubble dynamics was the bubble translation experiment of Vella-Coleiro and Tabor (1972), in which an isolated magnetic bubble domain is made to move by applying a magnetic field gradient. The experimental arrangement usually employed is shown in figure 1.11. Two conductors are laid, or deposited, upon the surface of the material. Both conductors carry a pulsed current in the same direction and if we imagine this direction to be into the paper, in figure 1.11, it is clear that a bubble of the polarity shown there will experience a force to the right. The experiment involves viewing the bubble domain in the microscope, applying a single pulse of sufficient ampli-

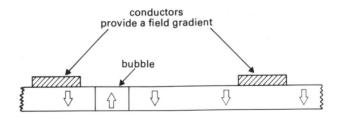

Figure 1.11 The bubble translation experiment

tude and duration to move the bubble through about one diameter and then calculating its velocity by assuming that the bubble actually moved that distance during the time of the pulse.

The difficulty with this experiment, as it is shown in figure 1.11, is that the mean value of the bias field is changed during the pulse except for the one particular case when the bubble domain is exactly between the two conductors. Vella-Coleiro and Tabor (1972) corrected for this by having two more conductors, not shown in figure 1.11, which produced a pulsed bias field to compensate for the change in mean bias field during the translation. A complete analysis of the correct way this compensation should be operated has been given by Gál (1975), who used the translation technique to study orthoferrites. A good example of the behaviour of bubbles in uncompensated translation can be found in the paper by Potter *et al.* (1975) which is also very interesting in that it gives the experimental details of using the translation technique on very small bubbles in amorphous GdCoCu films.

The bubble translation experiment at first seemed to show linear velocity against gradient field results, consistent with the simple wall mobility relationship given here as equation 1.2. The early results given by Vella-Coleiro (1972) covered a very wide range of different garnet compositions which had very different mobilities and few showed any very non-linear behaviour. This work coincided, however, with a very unexpected development in magnetic bubble domain work which was announced in four short notes in the *Bell System Technical Journal*, (Tabor *et al.,* 1972a, Wolfe and North, 1972, Bobeck *et al.,* 1972a and Rosencwaig, 1972), and in a paper by Malozemoff (1972). All five publications were concerned with the 'hard bubble', a bubble domain which appeared to be very much like the familiar bubble domain, as far as its appearance in the microscope went, until one tried to make it collapse by increasing the bias field. The hard bubble is so called because it is hard to collapse. One must increase the bias field to just about double the value needed to collapse ordinary bubbles before the hard bubble will disappear and the hard bubble diameter becomes so small before this collapse that a good microscope is needed to resolve the collapse diameter even in a garnet which supports normal bubbles of 5 μm diameter.

The other very unexpected experimental result which now came out (Tabor *et al.,* 1972a) was the behaviour of hard bubbles in the Vella-Coleiro and Tabor translation experiment. Instead of moving parallel to the applied magnetic field gradient, hard bubbles moved almost at right angles to this direction and, compared with the normal bubbles supported by the same film, they had a very low mobility.

Within a few months, a number of workers had repeated the hard bubble experiments and the experimental facts turned out to be far more complicated than expected. There were not just 'normal' bubbles and 'hard' bubbles by any means. There was found to be a whole spectrum of bubble 'states', from the point of view of the bubble translation experiment, in that bubbles could be found, in one and the same sample, which moved at discrete angles to the direction of the gradient, anywhere between —90° and +90°. The larger the deflection angle, the smaller the mobility of the bubble.

Two papers which gave the first details of these interesting results were by Voegeli and Calhoun (1973) and by Malozemoff (1973b). The discrete deflection angles and velocities observed by Voegeli and Calhoun (1973) are reproduced in figure 1.12. The material used was an LPE film of $(EuY)_3(GaFe)_5O_{12}$, 4.80 μm thick and all the results shown in figure 1.12 are for 5 μm bubbles moving in the same gradient field of 0.38×10^{-4} T/μm. The 'states' have been labelled with their state numbers, S, according to the theory of Voegeli and Calhoun (1973) which will be discussed in chapter 4. We need only remark here that the origin of the bubble deflection, and of the properties of hard bubbles, was attributed to some fine structure in the bubble wall and that the S-number is a simple way of describing this structure. The really fast bubble, labelled $S = +1$, was thought to be the one with the simplest wall structure, a wall having the form of figure 1.9c or d all the way around its perimeter. This conclusion followed from theoretical considerations (Slonczewski *et al.*, 1972) and the idea was supported by some more data given by Voegeli and Calhoun (1973) which is reproduced in figure 1.13 and shows the results of applying the bubble collapse experiment to bubbles whose state had been previously determined by bubble translation.

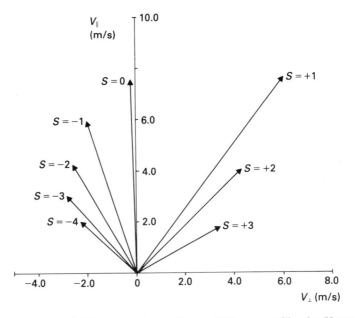

Figure 1.12 The bubble states observed in a LPE garnet film by Voegeli and Calhoun (1973). $V_{\parallel}$ is the component of velocity parallel to the applied gradient $V_{\perp}$ is the perpendicular component

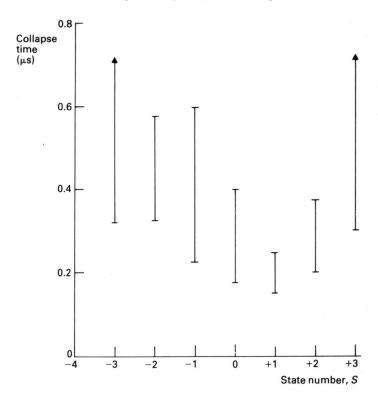

Figure 1.13 The bubble collapse data given by Voegeli and Calhoun (1973) for bubbles whose state had been previously identified by translation

As figure 1.13 shows, only the $S = +1$ bubble was found to have a well defined collapse time. Bubbles belonging to other states were more difficult to collapse and had a far greater statistical nature in their collapse process. The conclusion was that the $S = +1$ bubble had the simple wall structure of figure 1.9 c or d, while the other bubbles were thought to have a mixture of the two screw-senses, shown in figure 1.9c or d, alternating from one sense to the other as we proceed around the wall. Such a model was used to explain the high collapse field of hard bubbles and their deflection angles.

As more and more experimentalists took up the bubble translation experiment, the problem of the bubble states became more and more complex. Argyle *et al.* (1976a) described experiments in which a small modulation was superimposed upon the bias field in a translation experiment. This revealed three distinct kinds of bubbles which had all been previously identified as $S = +1$. When the bias field modulation was switched on, and an in-plane magnetic field was also applied, parallel to the direction of the conductors used to produce the pulsed gradient field, the $S = +1$ bubbles were observed to do one of three possible things: (a) nothing, (b) move to the left, perpendicular to the applied in-plane field, (c) move to the right.

Further experiments of this kind, described by Argyle *et al.* (1976c), revealed bubbles apparently having $S = +1/2$ which could move either parallel or antiparallel to the direction of the in-plane field when the bias field was modulated and, finally, four kinds of $S = -1/2$ bubbles were found which moved at $45°$, $135°$, $225°$ and $315°$ to the in-plane field. Bubble dynamics was beginning to compete with fundamental particle physics as a way of showing nature to be perverse.

1.10 The Application of High-speed Photography to Bubble Domain Dynamics

The use of high-speed photography in bubble domain dynamics was mentioned briefly in a long paper by Slonczewski *et al.* (1972) which dealt with the statics and dynamics of hard bubbles and gave the first tentative data and theory for the deflection angles we have discussed here in connection with figure 1.12. A year passed before the first high-speed photographs were published, by Morris and Malozemoff (1973), Malozemoff (1973a) and Zimmer *et al.* (1974a), showing the behaviour of initially straight walls and bubble domains when these were subjected to pulsed fields.

Interest became really intense when Malozemoff and DeLuca (1975) published their results of combining high-speed photography with the bubble translation experiment which was described in the previous section. The results showed that a great deal of the mystery surrounding this experiment came about because the bubble domain continued to move after the drive pulse was over. This unexpected overshoot of the moving bubble domain immediately explained the results obtained previously by Beaulieu and Voegeli (1974), in which bubble domains in translation appeared to overshoot what should have been their final position. The overshoot also explained the very high scatter which many workers had observed in their data from the bubble translation experiment. This followed because the bubble was found to move a well defined distance during the drive pulse but the amount of overshoot was found to be very variable (Malozemoff and Slonczewski, 1975). High-speed photography was also used to study the translation of really hard bubbles by Patterson *et al.* (1975) and similar overshoot was found in this case. It became clear that the experimentalists who were using bubble translation with an ordinary microscope, and thus saw only the initial and final bubble positions, were not measuring what they thought they were at all. The high-speed photography showed that the overshoot could account for a large part, even the major part, of the total trajectory unless very special conditions of high bias field and very low drive were used.

The important conclusion to be made from this work was that the bubble velocities which had been deduced from bubble translation were much too high because it had been assumed that the bubble travelled the whole distance between its initial position and its final position during the pulse. A plot of what was now thought to be the true velocity against drive field characteristic did not show the nearly linear characteristic of the old data at all; there was again an apparent saturation velocity. The same kind of thing had been seen in the bubble collapse data which was discussed in section 1.8. As we reported there, however, the bubble collapse experiment had also been shown to suffer from rather similar overshoot problems when the experiment was examined critically using high-speed photography (Humphrey, 1975).

Vella-Coleiro (1975) confirmed the overshoot results of Malozemoff and DeLuca (1975) using a low power pulsed argon laser as a light source and an image intensifier which made it possible to take single-shot photographs. All previous work on the high-speed photography of bubble domains in motion had been done using dye lasers, pumped by pulsed ultra-violet nitrogen lasers. Such a combination gave sufficient light for photographs to be taken directly (Humphrey, 1975).

Vella-Coleiro (1975) looked at high g-value garnet layers, that is, garnets in which the angular momentum coupled to the magnetisation is very much smaller than normal, and found that bubbles in translation within these materials showed no overshoot effect. Vella-Coleiro (1976a, 1976b) also found that the overshoot effect in normal garnets, with $g \approx 2$, disappeared when the drive pulse was below 100 ns. A beginning had been made in understanding the problems of bubble dynamics under these conditions and the theory behind this will be given in chapter 4, section 4.13. The model which will be introduced there will be one where the wall structure changes completely because of the bubble motion and the main support for such a model comes from the more recent results of high-speed microscopy. For example, Vella-Coleiro (1976c, 1976d) looked at the well defined and very simple dynamic experiment of an isolated bubble contracting under a pulsed bias field. Using a stroboscopic system which had a time resolution of 1 ns and could resolve a 0.1 μm change in the bubble diameter some remarkable fine structure in the wall motion could be observed. Fast initial motion was seen, followed. by a very brief, but quite distinct, reversal in the direction of wall motion.

More experimental facts came in from workers in the field of high-speed microscopy applied to bubble domain dynamics. Zimmer *et al.* (1974b) published photographs showing how a diffuse wall could develop around bubble and stripe domains when these were subjected to a large expanding magnetic field pulse in the presence of an in-plane field. Gál *et al.* (1975) published pictures showing how the magnetostatic instability of an expanding bubble domain developed in detail. Ju *et al.* (1976) used high-speed photography to look at the effect of a small bias field pulse upon the shape of a bubble domain which had just undergone translation. This experiment showed up the fine structure present within the bubble wall at different points around its circumference; an idea which is central to the theory of chapter 4.

Later work by Ju and Humphrey (1977a, 1977b) and Kleparskii *et al.*

(1977) showed how the fine structure within the wall could change during the translation of a bubble domain in a pulsed field gradient. What is very striking about these more recent papers involving high-speed microscopy is the excellent quality of the micrographs. Particularly fine examples are the pictures which show how a bubble domain may be formed by chopping a stripe domain with a bias field pulse (Gallagher and Humphrey, 1977) and how different bubble states may be identified by observing the change in bubble shape during an in-plane field pulse (Gallagher *et al.*, 1979). When it is applied to the detailed changes which take place in bubble shape and velocity during the cycle of a standard magnetic bubble domain device, high-speed microscopy becomes a very important analytical tool for the device engineer (Speriosu *et al.*, 1979, Gál and Humphrey, 1979).

2 The Landau – Lifshitz Equation

2.1 Introduction

The paper by Landau and Lifshitz (1935) was referred to in sections 1.2 and 1.4 of the previous chapter as one of the major theoretical steps in the history of ferromagnetodynamics. This chapter is concerned, almost entirely, with the results which have come from this paper and, in particular, with the equation which has become known as the Landau–Lifshitz equation.

Landau and Lifshitz were concerned with three problems.

(1) What was the atomic scale structure of the 180° domain wall, discussed in the previous chapter in connection with figure 1.4, which should separate the magnetic domains in a simple uniaxial magnetic material of infinite extent?

(2) What happened when the sample of uniaxial material was no longer assumed to be infinite and, consequently, had a large number of magnetic domains so that the external field energy was reduced?

(3) What was the effect of an externally applied magnetic field; first a constant field applied along the easy direction of magnetisation and then a field which varied sinusoidally with time?

It is interesting to note that the title of the paper, 'On the theory of the dispersion of magnetic permeability in ferromagnetic bodies', refers only to the last of the three problems listed above. This is the ferromagnetodynamic problem we are really concerned with here but it is very useful to look at the first problem before we move into dynamics. This is the magnetostatic problem of equilibrium domain wall structure and will introduce us to a number of fundamental and important points which are essential to the theory of ferromagnetodynamics.

2.2 The Problem of Magnetostatic Stability

Landau and Lifshitz (1935) first looked at the problem of magnetostatic stability in a ferromagnetic medium having one preferred direction of magnetisation and which was assumed to be of infinite extent. A single crystal of cobalt, with its strongly preferred hexagonal axis, was cited as an example.

What, then, is the distribution of the magnetisation, M, within such a medium? The first obvious solution is that M lies along the preferred direction everywhere. The magnetic field, B, within the medium is then given by

$$B = \mu_0 \ (H + M) \tag{2.1}$$

which is true in general. Because we have removed the surface divergence of M to infinity and have no volume divergence of M, under our assumption of uniform magnetisation, H is zero in this case and equation 2.1 tells us that $B = \mu_0 M$.

If we take the z-axis as the preferred direction of magnetisation, the trivial solution described above applies equally well for either sign of M. Landau and Lifshitz proceeded by considering the situation where one half of the medium was magnetised along the positive z-direction and the other half along the negative. They then proposed a particular kind of structure for the boundary between these two infinite half spaces and obtained the details of its structure by minimising an expression for the total energy of the system.

Let us repeat this calculation, not by looking at the energy but by looking at the equilibrium between the magnetisation and the total magnetic field at any point in the medium. In this way, we should end up with a partial differential equation for the problem of the distribution of magnetisation, as opposed to the integral equation which the energy minimisation approach yields. The method we are using is known as micromagnetics and good references are the two books by Brown (1962, 1963), the article by Döring (1966) and a paper by Kronmüller (1971).

The fundamental equation of micromagnetics states that the torque, given by the vector product of the magnetisation and the total magnetic field, must vanish at every point in a medium which is in magnetostatic equilibrium. That is

$$(M \times F) = 0 \tag{2.2}$$

where F is the total magnetic field which is made up from four separate parts

(1) The externally applied magnetic field.
(2) The magnetic field due to the magnetisation of the medium. This is always given by equation 2.1 and we shall refer to this part of F as the magnetostatic field.
(3) A magnetic field which represents the effect of anisotropy, the anisotropy field, B_a.
(4) A magnetic field which represents the effect of exchange, the exchange field, B_{ex}.

The externally applied field is the easiest of these to deal with, provided we define H in equation 2.1 as being a function of the magnetisation and the sample geometry only. The externally applied field is then the field in which the sample is immersed, that is, the field which is present in the absence of the sample.

The magnetostatic field is the most difficult part of F to deal with and for this reason it will be discussed last of all, when we are ready to look at the particular problem of domain wall structure which concerns us here.

Let us now look at what is meant by anisotropy and exchange and see if we can find a way of expressing these as magnetic fields which depend upon the local value of $M(x, y, z)$ and its partial derivatives. These can then be substituted into equation 2.2 to yield a partial differential equation in M, the solution of which should give the detailed structure of the wall.

2.3 The Anisotropy Field

A magnetic material is said to have an easy or preferred direction of magnetisation if a minimum energy state is obtained by having the magnetisation lying along this easy direction. In a uniaxial material, any deviation of M away from the easy direction by an angle θ results in an increase in the energy density which, in many cases, may be accurately represented by the expression $K_u \sin^2 \theta$. The anisotropy constant, K_u, has units J/m^3.

When the uniaxial anisotropy does have this simple form, it may be represented by means of an equivalent magnetic field, the anisotropy field, B_a, which is linearly related to M. We are only considering cases in which M has a constant magnitude, M_s, and can vary in direction. The effect of anisotropy may be represented very simply in such a case by proposing that, if M should become inclined by an angle θ to the easy axis, a magnetic field $\hat{B}_a \sin \theta$ appears perpendicular to the easy axis and opposed to the component of M in this same direction. This means that

$$- \tfrac{1}{2} (M \cdot B_a) = K_u \sin^2 \theta \tag{2.3}$$

and

$$\hat{B}_a = 2K_u/M_s \tag{2.4}$$

is the maximum value of the anisotropy field.

Taking the easy axis as the z-axis, the components of the anisotropy field are

$$(B_a)_x = -2K_u \ M_x/M_s^2$$

$$(B_a)_y = -2K_u \ M_y/M_s^2 \tag{2.5}$$

and

$$(B_a)_z = 0$$

and may now be substituted into equation 2.2 as part of F.

It has become very common practice to represent the anisotropy of a material by the dimensionless parameter $Q = 2K_u/\mu_0 M_s^2$. We may use this to simplify equations 2.5, writing the field components as $-\mu_0 Q M_x, -\mu_0 Q M_y, 0$. An even more useful representation has been introduced by Thiele (1973a) who uses Q as a matrix. In the representation used here this idea yields

$$(B_a)_i = \mu_0 Q_{ij} M_j \tag{2.6}$$

for the anisotropy field, where i and j take the values 1, 2, 3 to represent x, y, z. In the uniaxial case, with the easy direction the 3 direction, we have

$$Q_{ij} = Q \begin{bmatrix} -1 & 0 & 0 \\ 0 & -1 & 0 \\ 0 & 0 & 0 \end{bmatrix} \tag{2.7}$$

and more complicated anisotropic effects can be introduced very easily by using a more subtle matrix, Q_{ij}.

The effect of anisotropy is often represented by an equivalent magnetic field which lies along the easy direction, compared to being perpendicular to it as we have it here. Equations 2.5 are even simpler then in that we have only one; $(B_a)_z = \mu_0 Q M_z$. There is, of course, no difference between the two representations as they differ by only a constant scalar. The representation used here seems to be more natural in that the artificial anisotropy field is zero when M lies along the easy direction. The important point about both representations is that they assume that the anisotropy is represented accurately by $K_u \sin^2 \theta$. Most practical materials show more complicated anisotropy than this and the accurate representation of the anisotropy energy may need careful consideration (Ascher, 1966).

2.4 The Exchange Field

Following Kittel (1949) we treat the spins in the lattice of a ferromagnet as classical vectors and consider the exchange interaction as a potential energy term

$$V = (-2JS^2/M_s^2) \, M_p \cdot M_q \tag{2.8}$$

where J is the exchange energy integral, S is the spin and M_p/M_s and M_q/M_s are unit vectors in the directions of the spins on the pth and qth sites. Equation 2.8 then gives the exchange interaction between neighbouring spins as giving the minimum energy when these spins are parallel.

This is isotropic exchange and also implies that all the magnetic moments in the lattice have the same magnitude. It may be valid for simple ferromagnets and can be applied for ferrimagnets and weak ferromagnets if we assume that the antiferromagnetic coupling in these more complicated materials is very much stronger than the ferromagnetic coupling. Equation 2.8 may then be applied to the net magnetic moments of neighbouring magnetic unit cells. For many ferrimagnetic materials this seems to be a valid assumption but we must always bear the limitations of this simple model in mind. Many of the magnetic materials which have evolved for technical reasons, particularly the garnets used in magnetic bubble technology, are very complicated and contain several different magnetic ions. It is known that these more complicated materials may show magnetic phases, all having different arrangements, or canting angles, between their many sublattices.

(Clark and Callen, 1968, Zvezdin and Matveev, 1972, Grzhegorzhevskii and Pisarev, 1974). In addition, it is not generally realised that the magnetic structure of the garnets is by no means completely understood yet. Some recent work by Oudet (1974, 1979) has shown that the experimental values of M_s for quite well known garnets can be explained far more accurately when a model is used which is quite different to the conventionally accepted one. This work is strongly supported by some recent neutron diffraction work by Moon and Koehler (1977).

To return to equation 2.8 we must convert this expression into an equivalent field representation, as we did with the anisotropy energy, and get a field term to substitute into equation 2.2 as part of the total field F.

To do this, we expand M_p, the value of $M(x, y, z)$ at the lattice point p, in terms of the value it has at its neighbouring lattice points. We must consider a particular kind of lattice in order to do this and choose the simple body centred cubic one shown in figure 2.1. Here the p site is at the origin and has eight nearest neighbours, q_1 to q_8, at the points given by

$$x = a/2, \ y = -a/2, \ z = \pm a/2$$

$$x = -a/2, \ y = -a/2, \ z = \pm a/2$$

$$x = -a/2, \ y = a/2, \ z = \pm a/2 \tag{2.9}$$

$$x = a/2, \ y = a/2, \ z = \pm a/2$$

We now write the Taylor expansion for M_p in terms of M_q

$$M_p = \left[1 + \left(x_{qp} \frac{\partial}{\partial x} + y_{qp} \frac{\partial}{\partial y} + z_{qp} \frac{\partial}{\partial z} \right) + \frac{1}{2} \left(x_{qp}^2 \frac{\partial^2}{\partial x^2} \right. \right.$$

$$+ y_{qp}^2 \frac{\partial^2}{\partial y^2} + z_{qp}^2 \frac{\partial^2}{\partial z^2} + 2x_{qp}y_{qp} \frac{\partial^2}{\partial x \partial y} + 2x_{qp}z_{qp} \frac{\partial^2}{\partial x \partial z} \tag{2.10}$$

$$\left. \left. + 2y_{qp}z_{qp} \frac{\partial^2}{\partial y \partial z} \right) \right] M_q$$

where the operator in square brackets connects components of M_p and M_q in the usual way, and x_{qp}, y_{qp} and z_{qp} are the coordinates of q_1 to q_8, given by equations 2.9 because p was chosen as the origin.

Summing equation 2.10 over the eight values of q given by equation 2.9 we find that the terms involving the first and the mixed derivatives cancel out and

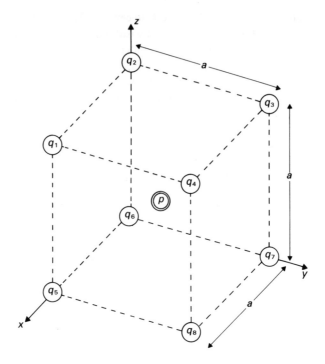

Figure 2.1 The body centred spin lattice for exchange calculations

we are left with

$$M_p = M_q + a^2 \left(\frac{\partial^2 M_q}{\partial x^2} + \frac{\partial^2 M_q}{\partial y^2} + \frac{\partial^2 M_q}{\partial z^2} \right) \tag{2.11}$$

Equation 2.8 for the exchange energy is thus

$$V = -2JS^2 \; [1 + a^2 \; (M \cdot \nabla^2 M)/M_s^2] \tag{2.12}$$

but we are not concerned with the constant part of this expression, only with the way in which it varies with **M**. We thus drop the constant term in equation 2.12 and also convert V to an energy density by dividing by a^3, the volume of the unit magnetic cell. This gives

$$E_{ex} = -[(2JS^2/a)/M_s^2] \; (M \cdot \nabla^2 M) \tag{2.13}$$

as the exchange energy density. We now represent this exchange energy density by introducing the equivalent exchange field, B_{ex}, such that

$$-\tfrac{1}{2} \, M \cdot B_{ex} = E_{ex} \tag{2.14}$$

From equations 2.13 and 2.14 it follows that

$$B_{ex} = (2A/M_s^2) \nabla^2 M \tag{2.15}$$

where we have adopted the usual convention of introducing the exchange constant $A = 2JS^2/a$. The constant, A, has the units of J/m and is typically between 10^{-11} and 10^{-12}.

2.5 The Magnetostatic Field

We have now obtained two of the results we need for substitution into equation 2.2. Equations 2.5 and 2.15 represent the effects of anisotropy and exchange in terms of M and, if these were the only contributions to F in equation 2.2, this would become a partial differential equation which might be soluble for M, subject to the correct boundary conditions.

We now have to consider the magnetostatic field. This part of F has already been described as the most difficult and the reason is that it has a completely different form to the anisotropy and exchange fields.

The magnetostatic field is given by equation 2.1

$$B = \mu_0 \, (H + M)$$

where H is given by the well known solution to the Poisson equation

$$H = (1/4\pi) \, \mathrm{grad} \left(\int_v \frac{\mathrm{div} \, M}{r_{ik}} \, dv + \int_s \frac{M \cdot n}{r_{ik}} \, ds \right) \tag{2.16}$$

A concise derivation of this result is given in the books by Sommerfeld (1950, 1952). The first integral in equation 2.16 is taken throughout the entire volume of the magnetised body, r_{ik} being the distance between the point of integration, i, which is surrounded by the volume element, dv, and the point k, at which H is to be evaluated. The second integral is taken over the entire surface of the magnetised body, n being a unit vector normal to the surface at the point of integration and directed inwards.

If equation 2.16 is now substituted into equation 2.1 in order to get an expression for the magnetostatic field, we end up with a result which contrasts sharply with our previous results for the anisotropy and exchange fields. The magnetostatic field depends upon the value of $M(x, y, z)$, and its derivatives, at every point in the magnetised body whereas the anisotropy and exchange fields, equations 2.5 and 2.15, only involve the local value of $M(x, y, z)$ and its derivatives. In general, substitution of equations 2.5, 2.15, 2.1 and 2.16 into equation 2.2 yields an integro-differential equation which is insoluble because the integral involved in equation 2.16 cannot be evaluated until the solution to the whole equation is known. This is a well known problem in domain theory and it has very profound implications for the dynamic problems which concern us here. Three recent papers by Aharoni (1975a, 1975b, 1975c) consider this in depth and give an excellent bibliography, while a fourth paper (Aharoni, 1976) considers the implications from the dynamic point of view.

It is because of these difficulties that we have postponed any detailed discussion of the magnetostatic field until we come to consider the particular wall structure, proposed by Landau and Lifshitz (1935), itself. These authors specifically chose a wall structure, or to be more exact they chose their boundary conditions, in such a way that equation 2.16 vanished and equation 2.12 became simply

$$B = \mu_0 M \qquad (2.17)$$

To obtain this result, Landau and Lifshitz (1935) considered an infinite sample, so that the second integral in equation 2.16 vanished. They also specified that M had constant amplitude in (x, z) and was a function of y only. This is the simplest vector field for which div $M = 0$ and so the first integral in equation 2.16 was also made to vanish. There are, of course, other possible forms for $M(x, y, z)$ which have the property div $M = 0$ but let us remain with the one chosen by Landau and Lifshitz.

2.6 The Structure of the Landau–Lifshitz Wall

To make the Landau–Lifshitz wall a little less artificial we can return to our discussion in chapter 1 concerning the contrast between the kind of domain walls expected in picture frame samples and in bubble domain materials. This was illustrated in figure 1.9 and the idea is repeated here in figure 2.2. This represents part of a picture frame sample, perhaps 1 mm thick and a few mm wide, and has a very narrow domain boundary between two domains which are both saturated along the z-axis, in opposite directions, this being the easy direction of magnetisation.

In such a situation, there is no normal component of M at the surface of the sample, which we must imagine as either infinitely long or as a closed frame, except at the very small area where the domain wall meets the surface. We have here a practical situation where the second integral in equation 2.16 would be negligible.

If we now invoke the second condition imposed by Landau and Lifshitz, that the magnetisation, M, rotates in (x, z) with constant amplitude as we go along the y-axis from one domain into the other. The first integral in equation 2.16 then vanishes because div $M = 0$.

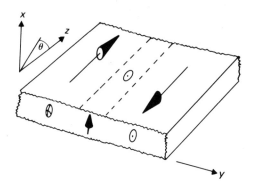

Figure 2.2 When the magnetisation lies in-plane within the domains there is no surface divergence of *M* except over the small region where the domain wall cuts the surface

Because *M* has only *x*- and *z*-components, so does the total field vector, *F*. These are given by

$$F_x = -(2K_u/M_s^2)\, M_x + (2A/M_s^2)\, \nabla^2\, M_x + \mu_0 M_x \qquad (2.18)$$

and

$$F_z = (2A/M_s^2)\, \nabla^2\, M_z + \mu_0 M_z \qquad (2.19)$$

which follow from equations 2.5, 2.15 and 2.17. Equation 2.2 thus has only a *y*-component

$$\frac{A}{M_s^2}\left(M_z\,\frac{d^2 M_x}{dy^2} - M_x\,\frac{d^2 M_z}{dy^2}\right) - \frac{K_u}{M_s^2}\, M_z M_x = 0 \qquad (2.20)$$

which is an ordinary differential equation because *M* is a function of *y* only. Because we assume that *M* has a constant magnitude, M_s, and simply rotates in (*x*, *z*) as we cross the wall, the components of *M* may be written

$$M_x = M_s \sin \theta \qquad (2.21)$$

$$M_z = M_s \cos \theta \qquad (2.22)$$

and substituted into equation 2.20 to give the very simple non-linear differential equation for θ

$$\frac{d^2\theta}{dy^2} - \frac{K_u}{A}\, \sin \theta \, \cos \theta = 0 \qquad (2.23)$$

To solve equation 2.23 we let $d\theta/dy = u$ and the variables separate in equation 2.23 to give

$$u \ du = (K_u/A) \sin \theta \cos \theta \ d\theta \qquad (2.24)$$

Equation 2.24 integrates directly to give

$$u^2 = (K_u/A) \sin^2 \theta + C_1 \qquad (2.25)$$

and if we choose $y = 0$ to be the central plane of the wall, the constant of integration, C_1, is zero. This follows because θ becomes constant at 0 or π, and $u = d\theta/dy$ tends to zero, as we move well into the magnetically saturated domains, $y \to \pm \infty$.

Taking the square root of equation 2.25 to get $u = d\theta/dy$, the variables again separate to give

$$\frac{(A/K_u)^{1/2} \ d\theta}{\sin \theta} = dy \qquad (2.26)$$

and equation 2.26 integrates directly to give

$$y = (A/K_u)^{1/2} \log_e (\pm \tan (\theta/2)) + C_2 \qquad (2.27)$$

The choice of sign in equation 2.27 indicates that the domain wall may have either a clockwise or an anticlockwise screw-sense associated with the rotation of M in (x, z). In other words, θ may equal $\pm\pi/2$ at $y = 0$. This latter boundary condition makes the constant of integration in equation 2.27 equal to zero and so the equation may be written

$$\pm \tan(\theta/2) = \exp(y/\Delta) \qquad (2.28)$$

where we have introduced the well known wall-width parameter

$$\Delta = (A/K_u)^{1/2} \qquad (2.29)$$

Finally, we may use the relationships between $\cos \theta$, $\sin \theta$ and the tangent of the half angle to combine equations 2.21 and 2.22 with equation 2.28 and obtain the very simple relationships

$$M_z = -M_s \tanh (y/\Delta) \qquad (2.30)$$

and

$$M_x = \pm M_s \ \text{sech} \ (y/\Delta) \qquad (2.31)$$

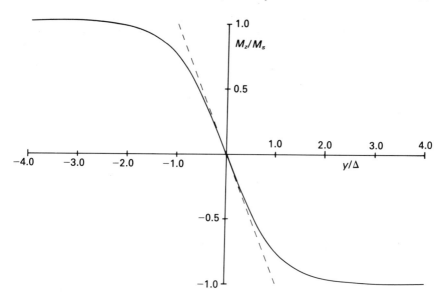

Figure 2.3 The variation of M_z/M_S with y/Δ for the Landau–Lifshitz wall

to define the static structure of the Landau–Lifshitz domain wall. Equation 2.30 has been plotted in figure 2.3 to illustrate that the Landau–Lifshitz wall occupies a width of about 2Δ and the rate of change of M_z, $\mathrm{d}M_z/\mathrm{d}y$, is $-M_S/\Delta$ over quite a distance about the centre of the wall.

We have thus arrived at an analytical description of the particular kind of domain wall discussed briefly in chapter 1 in connection with figure 1.4. It has been shown that this kind of solution could be expected in practice in the picture frame type of sample, shown in figure 2.2, where M lies predominantly in the plane of the sample. We can now go ahead and look at the theoretical model for the dynamic behaviour of this kind of wall and compare this with the experimental data. Before doing this, however, it is a good idea to ask what experimental evidence exists to confirm that the Landau–Lifshitz wall does really occur in nature. This can be done most easily by discussing some alternative wall structures.

2.7 Other Kinds of One-dimensional Wall Structure

The Landau–Lifshitz wall, described in the previous section is one-dimensional in the sense that the magnetisation is a function of y only. The wall has no in-plane structure variation.

There are two other kinds of simple one-dimensional wall structure which have been shown to be possible on theoretical grounds. The first one we shall describe briefly is originally due to Bloch (1932) and has been taken up recently by Bulaevskii and Ginzburg (1964, 1970). In this work the structure of the transition region between two domains of opposite directions of magnetisation is considered at a finite temperature, where the simple classical picture of the magnetisation as a constant vector is no longer strictly true. The results show that the Landau–Lifshitz wall structure should apply at temperatures well below the Curie point but that as we increase the temperature a transition should occur to a wall structure in which the magnitude of M is no longer constant. Instead of rotating in (x, z), as shown in figure 1.4, the magnetisation remains along the easy direction, z, decreases in magnitude to zero at the wall centre and then increases in the opposite direction to reach saturation again in the other domain.

Bulaevskii and Ginzburg (1964, 1970) showed that this kind of wall would be found in most ferromagnetic materials only in a small temperature range, from about 10 °C below the Curie point up to the Curie point itself. The problem would thus appear rather academic. In their later paper, however, they pointed out that in materials having a weak magnetisation and a high anisotropy, the Landau–Lifshitz solution would be found only at very low temperatures. The orthoferrite materials were cited as a particular example.

One very interesting point about the wall structure described by Bulaevskii and Ginzburg (1964, 1970) is that M_z is given by a function of exactly the same form as equation 2.30. Bloch (1932) proposed that the wall structure described by Bulaevskii and Ginzburg applied at all temperatures and, because the functional description of this wall looked identical to the wall described by Landau and Lifshitz this may be why the wall shown in figure 1.4 became universally called the 'Bloch wall'. The first use of this term, and the familiar diagram, appears in an article by Kersten (Becker, 1938).

The second kind of simple one-dimensional wall structure which has been proposed is called the 'Néel wall' (Néel, 1955). In this wall structure, M has constant magnitude and is only allowed to rotate in a plane perpendicular to the wall. Referring to figure 2.2, this is rotation of M in (y, z) and a wall of this kind would be expected in very thin films, so thin that the theoretical width of the Landau–Lifshitz wall would be of the same order as the film thickness. The 'Néel wall' would be expected in such thin films because it has no external magnetic field and this would add a significant amount to the total wall energy when the film is very thin. We shall consider a domain wall of very similar structure in chapter 3.

2.8 Experimental Evidence for the Landau–Lifshitz Structure

Let us now look at what experimental evidence there is for testing these various one-dimensional models. The Landau–Lifshitz wall is distinctly different to the other wall structures discussed above in that it produces an external magnetic field. This is clear from figure 2.2. There should be an external field, of the order of $\mu_0 M_s$, normal to the surface of the sample at the centre of the wall. This field should be found to take the two possible signs, belonging to the two possible screw-senses of the wall. It is also this field which is responsible for forming the

Bitter pattern, so the fact that Bitter patterns do form at all is some evidence in favour of the Landau–Lifshitz structure.

A more detailed look at the surface field was made by Krinchik and Verkhozin (1967), who used a polar Kerr effect apparatus of very high spatial resolution and were able to measure the width of the wall in a 1 mm thick single crystal of nickel. They found this to be consistent with the Landau–Lifshitz model in the case of nickel but not in the case of iron, where the wall appeared to be much too wide.

Still working on the surface, and exploiting the external field of the wall, it is possible to detect and measure the surface field with an electron beam which passes over the surface and is thus deflected by the field. This has been done by Shvetsov and Antip'yev (1973), who refer to the technique as glancing Lorentz microscopy. They were able to confirm that the external magnetic field from the walls in films of $NiFe_3$, between 0.1 μm and 0.3 μm thick, had the correct order of magnitude to be consistent with the Landau–Lifshitz model and could also take the two possible signs belonging to the two possible screw-senses.

There has been a great deal of far more indirect work on the problem of experimentally confirming wall structure. This is mainly concerned with a measurement of the wall energy. We shall not discuss this here because it mainly applies to the bubble domain kind of material in which the easy direction of magnetisation is perpendicular to the plane of the sample. These materials are considered in chapter 3. In any case, these very indirect techniques depend far too much upon an accurate knowledge of other material parameters and what we need is a truly microscopic look at what the magnetisation is doing inside the wall. Let us look at the experimental techniques which have been or might be used.

In order to find out what internal structure a domain wall really has, some kind of internal microscopic probe is needed. The only technique which has been developed so far is Lorentz microscopy, where electrons interact with the internal magnetic field of the wall. The problems of interpretation are considerable and are reviewed by Wohlleben (1967) who includes some excellent photographs in his paper. Unfortunately, the Lorentz technique is most suitable for very thin metal films and single crystal platelets, in which simple one-dimensional walls are not expected anyway. Most of the Lorentz work has been concerned with the much more complicated two-dimensional walls found in such thin samples. Harrison and Leaver (1973) describe work of this kind and give a good bibliography.

There are two other microscopic probe techniques which may be developed to give some information about the true structure of domain walls. The first is neutron diffraction, which must use very low energies and very sensitive detectors to work with the small field gradients and microscopic dimensions which the problem involves. Lermer and Steyerl (1976) have shown that the mean size of the domains in an iron film, and the mean width of the domain walls, could be determined from the very low energy neutron diffraction data and it is possible that these techniques will develop.

The other possibility is nuclear magnetic resonance (NMR). The work of Gossard *et al.* (1962) was cited by Bulaevskii and Ginzburg (1964) as some evidence for the kind of domain wall they proposed. The NMR data suggested that the magnetisation within the domain walls of $CrBr_3$ was lower than that in the domains themselves. Later work by Cobb *et al.* (1973) on $CrBr_3$, using more advanced NMR techniques, did not add any more information from the domain wall structure point of view and a great deal appears to be left for experimentalists to do in this field.

Our conclusion is that we are now going to look at the dynamics of the simple domain wall proposed by Landau and Lifshitz with only very indirect evidence that such a wall does, in fact, occur in magnetic materials.

2.9 The Formulation of the Landau–Lifshitz Equation

We can now consider the main subject of this chapter which is the most well known equation of ferromagnetodynamics. This is the equation given by Landau and Lifshitz (1935) which has been used by many authors as the starting point for theoretical work.

If we return to equation 2.2, the equation for magnetostatic equilibrium, it is clear that the dynamic, or non-equilibrium state in ferromagnetism will involve a situation in which the torque $(M \times F)$ is not zero. If there was no angular momentum associated with magnetism, this torque would simply cause M to move towards F so that they became parallel and equation 2.2 would be satisfied again. Things are not that simple, however, because the magnetisation of the material is due to spin and there is thus an angular momentum coupled to the magnetisation. In the simplest case, the angular momentum vector, Ω and the magnetisation, M, are related by a constant scalar, γ, which is usually called the magneto-mechanical ratio, so that

$$M = \gamma \Omega \qquad (2.32)$$

In the case of pure electron spin, $\gamma = -e/m = -1.76 \times 10^{11}$ (rad/s) per T. The angular momentum and the magnetic moment are antiparallel because the electron has a negative charge.

To return to the dynamic equation, the torque $(M \times F)$ must always be equal to the rate of change of angular momentum, $d\Omega/dt$. Using equation 2.32 we may write

$$\left(\frac{1}{\gamma} \right) \frac{dM}{dt} = M \times F \qquad (2.33)$$

as our first proposal for the fundamental equation of ferromagnetodynamics.

Equation 2.33 describes the uniform precession of the vector M about the total field F and does not represent the expected result that M and F should eventually become parallel to one another. For this reason, Landau and Lifshitz proposed the idea of adding a second term to the right hand side of equation 2.33 which had the direction $F - M$ and a magnitude which fell to zero when F and M became parallel. Such a vector may be formed as $(F - (F \cdot M) M/M_s^2)$ and the complete Landau-Lifshitz equation is written

$$\left(\frac{-1}{|\gamma|}\right) \frac{dM}{dt} = (M \times F) - \lambda \left[F - (F \cdot M) M/M_s^2\right] \qquad (2.34)$$

where λ is a positive constant having the same dimensions as M.

Equation 2.34 is the form in which the equation appeared in the original paper by Landau and Lifshitz (1935). We have used different symbols here and, in the original, square brackets were used to represent the vector product and parentheses for the scalar product. This was a very common notation in Europe at the time. We have also written the equation for materials with negative γ, which is the most common situation. There are materials with positive γ (LeCraw et al., 1975) and if we wrote equation 2.34 more generally, using γ instead of $-|\gamma|$, we would have to make λ negative when γ was positive. The final results would not depend upon the sign of γ, this simply determines the sense of the precession, and we shall avoid this complication.

Equation 2.34 represents M spiralling in towards F and eventually becoming parallel to F, whereupon both terms on the right hand side of the equation are zero. The second term which has been added is thus a damping term upon the previous purely precessional motion. It is interesting to note that a very similar damping term, involving a simple difference between the field and the magnetisation, was proposed by Kittel (1947) and later discarded (Kittel, 1948).

We do not find the complete Landau-Lifshitz equation being taken up in the literature until the papers by Beljers (1949) and Kittel (1950) were published. These papers really marked the beginning of the theoretical discussion of ferromagnetic resonance (FMR) which had been predicted in the Landau-Lifshitz paper but was not found experimentally until Griffiths (1946) published his results using electroplated nickel films at X and K band.

We next find the Landau-Lifshitz equation in a paper by Galt (1952) where it was written in the form

$$\frac{dM}{dt} = \gamma (M \times F) - \frac{\lambda}{M_s^2} (M \times M \times F) \qquad (2.35)$$

and it is this equation which is usually referred to as the Landau-Lifshitz equation. It does not occur in the paper by Landau and Lifshitz, however, and, although the same symbol, λ, is used for the damping term, this now has the dimensions of frequency, because it has been multiplied by γ. This has the advantage of making the equation apply for both signs of γ but the disadvantage of obscuring the condition $\lambda \ll M_s$ which Landau and Lifshitz underlined as essential to their original equation, 2.34. The condition $\lambda \ll M_s$ is implied by the Landau and

Lifshitz model in which the magnitude of M is constant and the damping comes from a very weak relativistic interaction. If λ no longer has the same dimensions as M a comparison of their magnitudes is obscure.

We do not enquire more deeply here into the atomic scale processes which may be responsible for the damping term in the Landau–Lifshitz equation. An excellent review of these problems has been given by Callen (1958) and by Sparks (1964) in his valuable book *Ferromagnetic Relaxation Theory*. Detailed descriptions of the damping mechanism for particular materials can be found in the papers by Harper and Teale (1967, 1969) and by Vella-Coleiro (1972) who give many important references to work in this field of solid state physics.

Gilbert and Kelly (1955) proposed a different form of equation 2.34

$$\frac{-1}{|\gamma|} \frac{\mathrm{d}M}{\mathrm{d}t} = M \times \left(F - \frac{\alpha}{|\gamma|M_s} \frac{\mathrm{d}M}{\mathrm{d}t} \right) \tag{2.36}$$

which follows directly from equation 2.34 if we use the vector triple product identity, introduce the dimensionless damping constant

$$\alpha = \lambda/M_s \tag{2.37}$$

and neglect terms in α^2. This is justified because of the original assumption of Landau and Lifshitz that $\lambda \ll M_s$.

In what follows we shall use the original Landau–Lifshitz equation, equation 2.34. The dimensionless damping parameter, α, will also be used, defined by equation 2.37.

2.10 The Application of the Landau–Lifshitz Equation to Domain Wall Motion

Landau and Lifshitz (1935) applied their equation, equation 2.34, to the problem of domain wall motion for the particularly simple case of the wall structure they had proposed. This is the structure given here by equations 2.30 and 2.31 and illustrated in figures 1.4 and 2.3. In order to avoid the complications of the magnetostatic field, they continued to consider the case of an infinite medium and also assumed that the wall structure remained almost identical to its previous static structure, when it was moving. In other words, they assumed a rigid wall structure.

There is a very important point here. If we assume that the wall moves as a rigid structure we have defined the space derivatives of M. They are already given by the differential equation, 2.23, which we solved to find this structure. If we now continue and assume that the wall is moving with a constant velocity, we have defined the time derivatives of M as well. We no longer have a differential equation to solve, the problem becomes purely algebraic, as we shall see, and can be solved by considering what is happening at any convenient point within the moving wall.

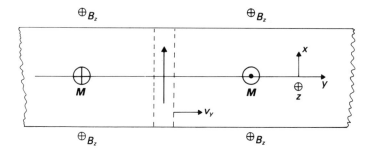

Figure 2.4 Coordinates used for the wall moving in an applied field B_z

The other assumption which follows, once a rigid wall model is adopted, is that the total magnetic field, F, previously given by equations 2.18 and 2.19, remains antiparallel to M in the moving wall as it was in the stationary wall. We then only need to consider the applied field, B_z, in our equations of motion. The situation is illustrated in figure 2.4 where we again emphasise the kind of sample of material we are dealing with. The magnetisation must lie in-plane and the sample thickness must be very much greater than the wall width. There is no question of applying the analysis given in this chapter to bubble domain materials or very thin films.

Let us now consider the sample, shown in figure 2.4, immersed in the applied field B_z. The domain on the left, in figure 2.4, is magnetised in the same direction as the applied field and consequently grows through the motion of the wall to the right with velocity v_y. Our diagram only shows M in the two domains and at the centre of the wall.

We now take the Landau-Lifshitz equation

$$\frac{-1}{|\gamma|} \frac{dM}{dt} = (M \times F) - \lambda [F - (F \cdot M) M/M_s^2] \qquad (2.38)$$

and, following the argument given above, we consider one convenient point within the wall. We choose the centre where $M_x = M_s, M_y = 0$. The total field, F, is now made up from the field belonging to the previous static solution, $M \times F = 0$, and therefore drops out of equation 2.38, and the applied field, B_z. Equation 2.38 thus has only two components

$$\frac{1}{|\gamma|} \frac{dM_y}{dt} = M_s B_z \qquad (2.39)$$

and

$$\frac{1}{|\gamma|} \frac{dM_z}{dt} = \lambda B_z \qquad (2.40)$$

Now, because $\lambda \ll M_s$ in the Landau–Lifshitz model we can concentrate on equation 2.39. This tells us that the proposal that the wall can move forward under the influence of the applied field B_z must involve the vector M developing a component M_y, a component in the direction of motion. It is not possible to assume that the wall is really rigid and that it maintains exactly the same form which it has when it is stationary.

As shown in figure 2.5, the magnetisation at the centre of the wall must tilt, to satisfy equation 2.39, by the angle ϕ shown. Let us go back to equation 2.38 again, with this modification, and see what happens.

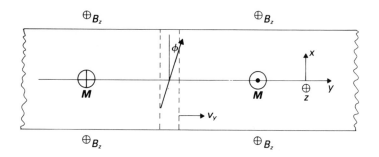

Figure 2.5 The moving wall must develop a component of M which lies along the direction of motion

Concentrating still on the centre of the wall, the vector M has components $M_x = M_s \cos \phi$, $M_y = M_s \sin \phi$ and $M_z = 0$. For the vector F, Landau and Lifshitz assumed that the parts of F due to exchange and anisotropy remained antiparallel to M, as they were when ϕ was equal to zero. This assumption made it possible to proceed with the calculation bringing in only the magnetostatic field and the applied field. They also assumed that M_z was still given by the function plotted in figure 2.3. We shall discuss the implications of these assumptions towards the end of this chapter; for the moment let us accept them.

If the sum of the exchange and anisotropy fields remains antiparallel to M, these do not enter into equation 2.38. The magnetostatic field is a particularly simple one to calculate in this case because the wall shown in figure 2.5 may be considered a thin layer with its magnetisation tilted out of the plane by the angle ϕ. There will then be no magnetic field due to the component of magnetisation normal to the plane, M_y, because equation 2.16 would give $H_y = -M_y$ for a thin layer so that equation 2.1 would give $B_y = 0$. In the plane of such a thin layer there is no such demagnetising effect and we have $B_x = \mu_0 M_s \cos \phi$.

We conclude that the components of the total field F which should be substituted into equation 2.38 for the case shown in figure 2.5 are

$$F_x = \mu_0 M_s \cos \phi \tag{2.41}$$

$$F_y = 0 \tag{2.42}$$

$$F_z = B_z \tag{2.43}$$

The components of equation 2.38 are now

$$-\frac{1}{|\gamma|} \frac{dM_x}{dt} = B_z M_s \sin \phi - \mu_0 \lambda M_s \cos \phi \sin^2 \phi \tag{2.44}$$

$$-\frac{1}{|\gamma|} \frac{dM_y}{dt} = -B_z M_s \cos \phi + \mu_0 \lambda M_s \cos^2 \phi \sin \phi \tag{2.45}$$

and

$$-\frac{1}{|\gamma|} \frac{dM_z}{dt} = -\mu_0 M_s^2 \cos \phi \sin \phi - \lambda B_z \tag{2.46}$$

We now consider the wall moving forward at constant velocity. Concentrating upon the centre of the wall, as before, M lies in (x, y) at the angle ϕ shown. As the wall moves forward, the vector M rotates about y and as it has constant magnitude, and ϕ is constant for constant velocity, equations 2.44 and 2.45 are both zero on the left hand side and both give

$$B_z = \mu_0 \lambda \cos \phi \sin \phi \tag{2.47}$$

The z-component of dM/dt, on the other hand is given by

$$\frac{dM_z}{dt} = M_s \dot{y} / \Delta \tag{2.48}$$

because $\partial M_z / \partial y = -M_s / \Delta$ at the centre of the Landau–Lifshitz wall, as shown in figure 2.3. This may be formally deduced from equation 2.30. Substituting equation 2.48 into equation 2.46 we obtain

$$v_y = (|\gamma| \Delta / M_s) (\lambda B_z + \mu_0 M_s^2 \sin \phi \cos \phi) \tag{2.49}$$

Equations 2.47 and 2.49 give us a picture of the domain wall moving forward at a constant velocity v_y, under the influence of a constant applied field B_z. Equation

2.47 shows that the effect of B_z is to make M tilt out of the plane of the wall by an angle

$$\phi = \tfrac{1}{2} \text{ arcsin } (2B_z/\mu_0\lambda) \tag{2.50}$$

where it comes into equilibrium with the total field and the effective field of the second term in the Landau–Lifshitz equation, which we could call the damping field. If we now substitute the equilibrium condition, equation 2.47 into equation 2.49 we obtain the relationship between the velocity and the applied field as

$$v_y = (|\gamma|M_s\Delta/\lambda) (1 + \lambda^2/M_s^2) B_z \tag{2.51}$$

As expected, the problem has become purely algebraic since we assumed a given wall structure and constant velocity.

Because the formulation of the Landau–Lifshitz equation demanded that $\lambda \ll M_s$ we can write equation 2.51 as

$$v_y = (|\gamma|M_s\Delta/\lambda) B_z \tag{2.52}$$

which is the often quoted result of Landau and Lifshitz, that the wall velocity is proportional to the applied field. Using equation 2.37 we can also write

$$v_y = (|\gamma|\Delta/\alpha) B_z \tag{2.53}$$

2.11 The Walker Limiting Velocity

The results in the previous section depended upon the assumption that the sum of the exchange and anisotropy fields remained antiparallel to M, as they are in the stationary wall, even though M was tilted out of the plane of the wall by the angle ϕ. It is tempting to see what happens if we assume that this is still true when ϕ becomes large and we are clearly well away from the original model of a domain wall in which M rotates about y and lies in (x, z).

Accepting this idea for the moment, equation 2.47 shows that the condition

$$B_z \leqslant \mu_0\lambda/2 \tag{2.54}$$

must be satisfied if equation 2.47 is to have a real solution for ϕ. When $B_z = \mu_0\lambda/2$, $\phi = \pi/4$. If we substitute $B_z = \mu_0\lambda/2$ into equation 2.52 we obtain a maximum or limiting wall velocity

$$v_W = |\gamma|\mu_0 M_s\Delta/2 \tag{2.55}$$

and this has become known as the Walker limiting velocity (Schryer and Walker, 1974).

It is certainly rather difficult to believe that the sum of the exchange and anisotropy fields is going to remain antiparallel to M, as it is in the stationary wall, when the angle ϕ becomes as large as $\pi/4$. The result is illuminating, however, because it directs our attention towards the real origin of the wall motion which is the very large magnetic field, $\mu_0 M_x$, in the plane of the wall. This acts upon the y-component of M in the z-component of $(M \times B)$ which is $(M_x B_y - M_y B_x)$. The product $M_y B_x$ becomes $\mu_0 M_s^2 \cos\phi \sin\phi$ and produces the rotation of M around y. The only part played by the applied field is to increase ϕ.

It should be mentioned that the alternative interpretation, that the origin of the wall motion is the so-called demagnetising field in the y-direction, is quite valid. This involves introducing the magnetic excitation vector H and arguing that because

$$B = \mu_0 \ (H + M) \tag{2.56}$$

in general, it is also generally true that

$$(M \times B) \equiv \mu_0 \ (M \times H) \tag{2.57}$$

We have adopted the practice of using the magnetic field vector B as the cause, or independent variable, of the effects we shall be describing. It is not necessary to bring the vector H into a problem in which M has to be used. Because of equation 2.56 only two of the three vectors need to be specified. Equation 2.56 must be satisfied everywhere, H is simply the vacuum polarisation in which the material polarisation, M, is embedded. There is a practical point about using B instead of H as well. The units of B, T, are simply related to the old units, Gauss and Oersted, by the factor 10^{-4}.

2.12 The Relationship to Ferromagnetic Resonance

The Landau–Lifshitz model has been shown to predict a linear relationship between the domain wall velocity and the applied field, equation 2.52, and a limiting velocity, equation 2.55. The early experiments, described in chapter 1, certainly suggested that a linear velocity–field relationship applied in a number of cases but any attempt to show that this behaviour was really due to the kind of precessional motion described here had to wait for the single crystal samples of the insulating magnetic materials, the ferrites and garnets, to become available.

The reason for this was that the damping constant, λ, in the Landau–Lifshitz equation could only be determined experimentally by ferromagnetic resonance on single crystal sphere shaped samples. Once it could be determined, by a resonance experiment, it seemed sensible to attempt to fit the kind of experimental data on wall motion, which had been obtained using the picture frame samples described in section 1.4, to the Landau–Lifshitz model.

In this section we shall describe the FMR experiment briefly and obtain the relationships between the material parameters and the field, frequency and line-width observed.

In the FMR experiment, a highly polished spherical sample of magnetic material is placed in a constant magnetic field, B_z, and a high frequency magnetic field, B_y. The magnitude of B_y is much smaller than B_z. The diameter of the sample is much smaller than the wavelength of the high frequency excitation.

We shall restrict our discussion to a uniaxial magnetic material in which the constant field, B_z, is directed along the easy direction of magnetisation. Extending the argument to other directions, or crystals of higher symmetry, does not add any fundamentally new behaviour.

Let us now describe this simple situation by means of the Landau–Lifshitz equation

$$\frac{-1}{|\gamma|} \frac{dM}{dt} = (M \times F) - \lambda \left(F - (F \cdot M) M/M_s^2 \right) \qquad (2.58)$$

The total magnetic field, F, is made up from the constant field B_z, the small high frequency field, B_y, and the anisotropy field. Thus, from equation 2.5

$$F_x = -\mu_0 Q M_x$$

$$F_y = B_y - \mu_0 Q M_y \qquad (2.59)$$

$$F_z = B_z$$

where

$$Q = 2K_u/\mu_0 M_s^2 \qquad (2.60)$$

There is no contribution from the exchange field in this case because we are considering the uniform precession mode so that M is a function of time only. We shall discuss the conditions which are essential to ensure uniform precession in a moment. There is also no need to include the magnetostatic field in this problem because, in a spherical sample, this is given by $2\mu_0 M/3$, which is always in the same direction as M and will drop out of equation 2.58.

Let us now consider precession at a very small angle about the z-axis, along which M is constrained by the large applied field, B_z. This means that we can write

$$M_x = \hat{M}_x e^{j\omega t}$$

$$M_y = \hat{M}_y e^{j\omega t}$$

$$M_z = M_s \qquad (2.61)$$

$$B_y = \hat{B}_y e^{j\omega t}$$

and substitute equations 2.59 and 2.61 into equation 2.58 to obtain the x-component

$$-j(\omega/|\gamma|)\ \hat{M}_x = [\hat{M}_y + (\lambda/M_s)\ \hat{M}_x]\ (B_z + \mu_0 Q M_s) - M_s \hat{B}_y \qquad (2.62)$$

and the y-component

$$-j(\omega/|\gamma|)\ \hat{M}_y = [(\lambda/M_s)\ \hat{M}_y - \hat{M}_x]\ (B_z + \mu_0 Q M_s) \qquad (2.63)$$

where we have neglected second order terms. The z-component of equation 2.58 vanishes because of our simplification that $M_z = M_s$.

From equations 2.62 and 2.63 we obtain the resonance equation, neglecting terms in λ^2/M_s^2, as

$$\hat{M}_y = \frac{\hat{B}_y M_s\ [(\lambda/M_s)\ (B_z + \mu_0 Q M_s) - j(\omega/|\gamma|)]}{[(\omega/|\gamma|)^2 - (B_z + \mu_0 Q M_s)^2] - 2j(\omega/|\gamma|)\ (\lambda/M_s)\ (B_z + \mu_0 Q M_s)} \qquad (2.64)$$

This predicts ferromagnetic resonance at a frequency

$$\omega_0 = |\gamma|\ (B_z + \mu_0 Q M_s) \qquad (2.65)$$

and a line-width, in terms of frequency, of

$$\delta\omega = (2\lambda\omega_0/M_s) \qquad (2.66)$$

or in terms of magnetic field

$$\delta B_z = (2\omega_0 \lambda/|\gamma|M_s) \qquad (2.67)$$

where, by line-width, we mean the difference between the off resonance conditions which reduce the value of $\hat{M}_y$ to $1/\sqrt{2}$ of its value at resonance. This is defined specifically in figure 2.6. The experiment is usually done at fixed frequency and B_z is varied because then the sample can be mounted in a very low loss resonant cavity and the $1/\sqrt{2}$ condition determined accurately by finding the two values of B_z at which the total power absorbed by the sample is one half of the power absorbed at resonance. In this way, the value of $\hat{B}_y$ can be kept constant.

The conditions for this uniform precession mode of ferromagnetic resonance require very high spatial uniformity of the two fields, B_z and $\hat{B}_y$. This implies the use of a spherical sample, which is most important because values of λ/M_s have been quoted in the literature which are deduced from FMR on thin films or plates. It is not possible to excite only the uniform precession mode in such samples. The non-uniform modes of precession, usually referred to as the magnetostatic modes (Walker, 1957) show considerably greater line broadening than the uniform precession mode.

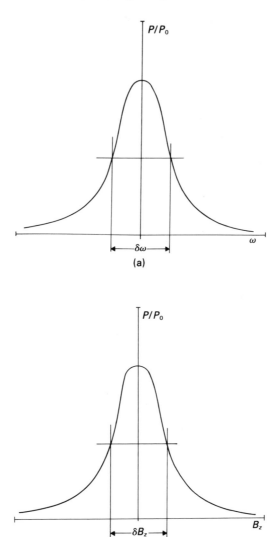

Figure 2.6 In the FMR experiment the power absorbed by the sample is measured, either at constant field as a function of ω, as in (a), or, more often, at constant frequency as a function of field, (b)

The uniform precession mode also breaks down, and higher modes are excited, if the power level is too high. With very low loss materials, like the garnets, this can happen even though $\hat{B}_y \ll B_z$ and a number of examples and useful data have been collected by von Aulock (1965). A further problem with FMR measurements in very low loss materials is that the surface of the sample can act as a source of excitation for higher order modes. Experiments by LeCraw *et al.* (1958) and Spencer *et al.* (1959) on yttrium iron garnet (YIG) showed this dramatically. The line-width for a spherical sample could be reduced by over an order of magnitude by correct polishing.

2.13 Experimental Work; the Magnitude of λ and its Implications

It should be emphasised again that the λ we are using here, equations 2.66 and 2.67, is the damping constant introduced by Landau and Lifshitz (1935) which has the dimensions of magnetisation. The symbol λ is often used in the literature of ferromagnetodynamics to represent $\gamma\lambda$, which has the dimensions of frequency. The reason for this confusion was the way in which Galt (1952) wrote his form of the Landau–Lifshitz equation and was discussed here in connection with equation 2.35.

No confusion in symbolism will arise if we use

$$\alpha = \lambda/M_s \tag{2.68}$$

in what follows because this dimensionless damping parameter is quite common in the literature and, of course, does not depend upon the system of units. The most common expression for the experimental determination of α, from FMR, is

$$\alpha = (\delta B_z)\,(|\gamma|/2\omega_0) \tag{2.69}$$

which follows from equation 2.67.

An essential assumption of the Landau–Lifshitz model is that M has constant magnitude. From this it follows that $\lambda \ll M_s$ or $\alpha \ll 1$. Let us now look at some FMR data and make a comparison with the prediction of the Landau–Lifshitz equation that the wall mobility, originally defined in equation 1.2 as

$$v_n = \mu_w(B - B_0) \tag{2.70}$$

is given by

$$\mu_w = (|\gamma|\mathit{\Delta}/\alpha) \tag{2.71}$$

from equation 2.53, when we assume that the coercivity B_0 simply subtracts from the applied field, B. We also have the condition that the inequality

$$(B/\mu_0 M_s) \ll (\alpha/2) \tag{2.72}$$

which follows from equations 2.54 and 2.68, must be satisfied if this simple mobility model is to apply.

Let us now look back at chapter 1 and consider the experimental results given there for nickel ferrite and for YIG.

The data given by Galt (1954) for nickel ferrite, shown in figure 1.6, suggest a value of $\mu_w = 2.6 \times 10^6$ (m/s) per T for applied fields below 0.3×10^{-4} T and 201 K. The FMR data given by von Aulock (1965), after correcting for the eddy-current contribution which is significant in the microwave resonance experiment but not in the wall mobility experiment, suggests that $\alpha \approx 4 \times 10^{-3}$. The value of $\mu_0 M_s$ for nickel ferrite is 0.42 T, all this data being for 201 K. The inequality equation 2.72, then tells us that the applied field in the mobility experiment must be less than 8×10^{-4} T and our first conclusion is that Galt's experiments were well within the range of applied field for which the Landau–Lifshitz model applies.

Using Galt's values for $A = 1.09 \times 10^{-11}$ J/m, $K_u = 5.4 \times 10^3$ J/m³ and $|\gamma| = 1.9 \times 10^{11}$ (rad/s) per T, equations 2.29 and 2.71 predict a value of $\mu_w = 2 \times 10^6$ compared to the measured value of 2.6×10^6. We shall comment in some detail about the fact that we have treated this apparently cubic magnetic material as though it is uniaxial at the end of this chapter.

The conclusion is that Galt's experiments on nickel ferrite gave support to the Landau–Lifshitz model. So did the later experiments on manganese ferrite by Dillon and Earl (1959) and on nickel by Leaver and Vodjani (1970). In all these experiments, the fields used were well below the critical value given by equation 2.72 and the results followed equation 2.70 with a small positive value of B_0.

Let us now consider the other experiments discussed in chapter 1 which are relevant to the situation being discussed here. The results for YIG (Harper and Teale, 1969) were shown in figure 1.7 and at 295 K these suggest a mobility of 3×10^5 (m/s) per T, with a small positive intercept, $B_0 \approx 0.02 \times 10^{-4}$ T. Now the FMR line-width in YIG is very small and for a really perfect surface polish (Spencer *et al.*, 1959) $\alpha \approx 10^{-4}$. The value of $\mu_0 M_s$ is 0.17 T at room temperature so that equation 2.72 tells us that the applied field in the wall mobility experiment must be less than 0.085×10^{-4} T.

The results for YIG, shown in figure 1.7, are all at applied fields greater than 0.085×10^{-4} T. It is rather striking that they happen to begin at precisely this value. It is certainly not surprising that the measured wall mobility bears no resemblance to the value predicted by equation 2.71. Typical values of $A = 4.3 \times 10^{-12}$ J/m, $K_u = 580$ J/m³ and $|\gamma| = 1.76 \times 10^{11}$ (rad/s) per T, for YIG, predict $\mu_w = 1.5 \times 10^8$ (m/s) per T; nearly three orders of magnitude greater than the experimental result.

We may conclude that there is some evidence in support of the Landau–Lifshitz model when the experimental situation is the simple one shown in figure 2.2 and the applied fields are small enough to satisfy equation 2.72. The remark above concerning figure 2.2 must be emphasised. We should only expect the behaviour predicted by Landau and Lifshitz when we have in-plane magnetisation and a sample much thicker than the domain wall width. It is not easy to have this ideal situation with many materials because the available crystals are too small.

Vella-Coleiro *et al.* (1971, 1972c) used crystals, cut from flux grown rare-earth iron garnets, typically 3 x 4 x 1 mm. The easy direction of magnetisation was normal to the plane of these single crystal plates, the very opposite of our figure 2.2. The Landau–Lifshitz wall structure might be expected to apply to quite a large part of the domain walls in these samples, because they are so thick compared to the wall width, but surface effects would be very significant. Very good agreement with the Landau–Lifshitz model was claimed by Vella-Coleiro *et al.* (1972c) for some of these crystals when the mobility was deduced from measurements of the high frequency susceptibility. The values of λ/M_s obtained, however, were either of the order of unity or exceeded unity, which is completely incompatible with the model. In earlier experiments by Vella-Coleiro *et al.* (1972a) the same experimental method was used on small crystals of yttrium gadolinium iron garnet, having a wide range of composition. The mobilities deduced from high frequency susceptibility measurements in this case were much lower than those predicted by the Landau–Lifshitz model using the FMR data. This may have been due to the kind of sample used or because the applied field was not kept below the critical $\mu_0 \lambda/2$.

2.14 Wall Acceleration; the Döring Mass

So far, we have only considered the Landau–Lifshitz wall structure moving at constant velocity and in a constant applied field. Any transient effects have been tacitly assumed to be well damped.

This steady-state condition was also assumed by Landau and Lifshitz (1935) in their original treatment of the problem, until they came to the final section of their paper in which they considered the case of applied fields which varied sinusoidally with time. For the case which concerns us here, Landau and Lifshitz introduced no additional terms into their equation of motion, equation 2.34 but simply substituted a sinusoidally varying field into their result for the velocity–field relationship, equation 2.52. They concluded their paper by considering the effect of a field applied perpendicular to the plane of their wall structure, this field varying sinusoidally with time, and this must have been the first theoretical consideration of ferromagnetic resonance.

A more detailed analysis of the problem of an accelerating wall was given much later by Döring (1948) who considered the possibility that the magnetic energy of the wall might change as the velocity changed. His conclusion was that there must be an effective mass associated with a domain wall which has the simple Landau–Lifshitz structure.

Döring's analysis was very clearly summarised in a note by Becker (1952) and can be understood by reference to figures 2.4 and 2.5. When the wall is stationary, as in figure 2.4, there is a large magnetic field at its centre $\mu_0 M_s$, directed along x. The energy density at this point, due to this magnetostatic field, is $-\frac{1}{2}\mu_0 M_s^2$. There is, of course, a much larger positive energy density associated with the wall due to exchange and anisotropy but, as we emphasised in sections 2.9 and

2.10, the Landau–Lifshitz model assumes that the wall moves without any change in its width or the way in which M_z varies with y. The sum of the exchange and anisotropy field is assumed to remain antiparallel to M. This is why these fields do not come into the equations of motion, 2.44, 2.45 and 2.46.

Now let us consider what happens when the wall moves. As shown in figure 2.5, M must take up an angle ϕ to the wall plane. The magnetic field at the wall centre changes from $\mu_0 M_s$ to $\mu_0 M_s \cos \phi$, directed along x. There is no field along y. This means that the magnetostatic energy density has increased by

$$\delta E = \tfrac{1}{2} \, \mu_0 M_s^2 \, \sin^2 \phi \qquad (2.73)$$

We now take equations 2.47 and 2.52 to obtain a relationship between the velocity of the wall and the angle ϕ as

$$\cos \phi \, \sin \phi = v_y / \mu_0 |\gamma| M_s \Delta \qquad (2.74)$$

but consider only small values of ϕ, that is velocities well below the Walker limiting velocity given by equation 2.55, so that $\sin \phi \approx \phi$, $\cos \phi \approx 1$. We can then substitute equation 2.74 into equation 2.73 to get

$$\delta E = v_y^2 / 2\mu_0 \gamma^2 \Delta^2 \qquad (2.75)$$

If we convert this energy density into an energy per unit wall area, by multiplying by the effective wall width 2Δ, shown in figure 2.3, we can say that

$$E = \tfrac{1}{2} \left(\frac{2}{\mu_0 \gamma^2 \Delta} \right) v_y^2 \qquad (2.76)$$

is the energy per unit area of a domain wall due to its velocity. Equation 2.76 is written in the particular way it is to bring out the analogy to kinetic energy, $\tfrac{1}{2} m v^2$. From such a view point we can say that the domain wall has an effective mass per unit area, in kg/m^2

$$m_D = 2/\mu_0 \gamma^2 \Delta \qquad (2.77)$$

Equation 2.77 is identical to the one given by Döring (1948) and Becker (1952). Their expression, in cgs units, is $(2\pi \gamma^2 \Delta)^{-1}$ g/cm^2.

It should be noted that the energy density, equation 2.75, was converted into an energy per unit area, equation 2.76, simply by multiplying by 2Δ. This is a fair enough approximation because we are really considering the energy density of the in-plane field of the wall, not just the component along the x-direction.

2.15 A Field Interpretation of the Effective Wall Mass

We have arrived at the concept of an effective wall mass by using a rather mechanical analogy. This tends to cover up the real nature of the effect because a moving domain wall is not a material particle in any sense and all the forces which act upon it have a magnetic origin. The inertial effect, which Döring (1948) and Becker (1952) showed to follow from considering the energy of the wall and Lagrange's equations of motion, should follow equally well if we consider only the field equations.

To do this, we return to the three components of the Landau–Lifshitz equation, which were written out as equations 2.44, 2.45 and 2.46. When these were used to consider motion at a constant velocity, the left-hand sides of equations 2.44 and 2.45 were set equal to zero. When the angle ϕ, in figure 2.5, is changing with time this is no longer true.

In general, the y component of dM/dt is given by $(M_s \cos \phi)\, d\phi/dt$ at the centre of the moving wall and we can substitute this into equation 2.45, cancel $M_s \cos \phi$ on either side and obtain

$$\frac{-1}{|\gamma|}\frac{d\phi}{dt} = -B_z + \mu_0 \lambda \cos \phi \sin \phi \qquad (2.78)$$

Equation 2.48, for the z-component of dM/dt, is still true in general because we have defined θ as the angle which M makes with the z-axis. It follows that equation 2.49 is unchanged when ϕ is a function of time. For small values of ϕ, equation 2.49 may be rearranged to give

$$\phi = (v_y/\mu_0|\gamma|M_s\varDelta) - (\lambda B_z/\mu_0 M_s^2) \qquad (2.79)$$

Equation 2.79 is now substituted into equation 2.78, ϕ again being assumed small, to give the differential equation

$$\frac{1}{\mu_0|\gamma|\lambda}\frac{dv_y}{dt} + v_y = \frac{|\gamma|M_s\varDelta}{\lambda}\left(1+\frac{\lambda^2}{M_s^2}\right)B_z + \frac{\varDelta}{\mu_0 M_s}\frac{dB_z}{dt} \qquad (2.80)$$

For constant applied field, equation 2.80 yields the same steady state velocity

$$v_y = \frac{|\gamma|M_s\varDelta}{\lambda}\left(1+\frac{\lambda^2}{M_s^2}\right)B_z \qquad (2.81)$$

as we obtained before as equation 2.51.

To show that equation 2.80 is identical to the mechanical analogy, used in the previous section, we consider the case of a constant applied field and neglect the term λ^2/M_s^2. Equation 2.80 is then

$$\frac{1}{\mu_0|\gamma|\lambda}\frac{dv_y}{dt} + v_y = \frac{|\gamma|M_s\Delta}{\lambda}B_z \qquad (2.82)$$

The force per unit area on a domain wall of this kind is simply $2M_sB_z$ so we multiply both sides of equation 2.82 by $2\lambda/|\gamma|\Delta$ and write

$$\frac{2}{\mu_0\gamma^2\Delta}\frac{dv_y}{dt} + \frac{2\lambda}{|\gamma|\Delta}v_y = 2M_sB_z \qquad (2.83)$$

which is in direct analogy to the equation

$$m\frac{dv}{dt} + \beta v = F \qquad (2.84)$$

for a particle in a viscous fluid. Comparing equations 2.83 and 2.84 shows that the mass is given by equation 2.77 again and the viscous force per unit area is the familiar $2M_s/\mu_w$, where μ_w is given by equation 2.71.

To go back to equation 2.80, we can obtain an idea of the experimental implications of the wall mass. As we have made the equation of motion linear by considering only very small values of ϕ, which means applied fields very much smaller than $\mu_0\lambda/2$, it is useful to consider the strictly linear response of the domain wall to an oscillating field, $B_z = \hat{B}_z e^{j\omega t}$. The wall velocity will then be $v_y = \hat{v}_y e^{j\omega t}$, where $\hat{v}_y$ is now complex. Equation 2.80 gives the relationship

$$\hat{v}_y = \mu_w\hat{B}_z\ [1 + j(\omega/\omega_1)]/[1 + j(\omega/\omega_2)] \qquad (2.85)$$

where

$$\omega_1 = |\gamma|\mu_0M_s/\alpha = \omega_s/\alpha \qquad (2.86)$$

$$\omega_2 = \alpha|\gamma|\mu_0M_s = \alpha\omega_s \qquad (2.87)$$

The frequency $\omega_s = |\gamma|\mu_0M_s$ is what we might call the free precession frequency of a magnetic material and is usually found in the microwave region. For example; in YIG, where $|\gamma| = 1.76 \times 10^{11}$ (rad/s) per T and $\mu_0M_s = 0.17$ T at room temperature, we have $\omega_s = 4760$ MHz. Because $\alpha \approx 10^{-4}$ in this material, the frequency ω_1 is too high to be of any practical importance and we may simplify equation 2.85 to

$$\hat{v}_y = \mu_w\hat{B}_z/[1 + j(\omega/\omega_2)] \qquad (2.88)$$

The frequency ω_2 in YIG would be of the order of 500 kHz, according to equation 2.87. Above this frequency, the wall inertia would cause the wall to lag behind the applied field. Such effects would be difficult to see in YIG for the reasons discussed in section 2.12; the applied field must be below $\mu_0\lambda/2$, which is less than 10^{-5} T, for this model to apply anyway. In other materials, however, experimental work does support the idea of domain wall inertia and this will be reviewed in the next section.

2.16 Experimental Work on Wall Inertia

Attempts to observe the effects of domain wall inertia have been made by measuring the way in which the magnetic susceptibility of a sample depends upon frequency. Equation 2.80 was for the case of an isolated domain wall in an infinite sample. When a finite sample is involved a magnetostatic field term must be included in B_z which depends, in the case of a single wall, on the sample dimensions. For a sample with many walls, the magnetostatic field depends upon the size of the domains and the sample dimensions.

In all cases the magnetostatic field may be represented by a term linear in a small displacement of the wall and acting as a restoring force upon the domain wall. The addition of this term to equation 2.80 makes it second order in wall position and predicts a resonance in the wall motion. The resonant frequency depends upon the wall mass and the stiffness of the restoring force. The sharpness of the wall resonance depends upon the wall mobility and if the mobility is low, the viscous damping may then be predominant and no wall resonance will be seen.

Using this technique, Perekalina *et al.* (1961) saw a clear wall resonance at 360 MHz in a picture frame sample of single crystal cobalt ferrite. The expected resonant frequency was 500 MHz so that, in view of the uncertainty of some of the material constants agreement was quite good.

The experiments of Vella-Coleiro *et al.* (1972a), which were referred to in section 2.13, gave very good agreement between the measured and theoretical values of effective wall mass in a series of crystals having different compositions in the yttrium gadolinium iron garnet system. The fact that the wall damping in these samples was so much higher than expected was probably due to the magnetisation being out of plane, so that the wall structure is quite different to the Landau–Lifshitz structure near the surface of the sample, a point which is central to chapter 3 here. Increased damping, however, does not shift the position of the wall resonance very much and the effective wall mass is determined from this.

Finally, some experiments have been reported involving wall resonance in the presence of in-plane magnetic fields. Because the majority of these experiments have been done using bubble domain materials we shall postpone a discussion of the results to chapter 3.

2.17 Including the Exchange and Anisotropy Fields

We have now covered the most well known results which follow from the Landau–Lifshitz equation. These are, the linear velocity–applied field relation-

ship given by equation 2.53, the limiting velocity, with its associated field, given
by equations 2.55 and 2.54 and the concept of an eqivalent wall mass given
by equation 2.77.

In all these cases we have simply ignored the exchange and anisotropy fields.
It was pointed out in section 2.10 that the justification for this assumption
comes from the fact that we have treated the moving wall as a rigid structure,
having the same form when moving as when it is stationary. Because the sum of
the anisotropy and exchange fields is antiparallel to M when the wall is stationary
we assume that it is still antiparallel when the wall is moving. The exchange and
anisotropy fields drop out of the Landau–Lifshitz equation in such circumstances
and this complicated non-linear partial differential equation becomes a non-
linear ordinary differential equation. We went even further in our simplification,
in fact, and assumed a constant velocity. This reduced the problem even further
because it is then simply algebraic, as shown by the concluding equations of
section 2.10.

Such a drastic simplification depends upon two points. In the first place we
must show that the wall structure is difficult to change when the angle ϕ, first introduced in connection with figure 2.5, is small. We could then accept the assumption that the sum of the exchange and anisotropy remains antiparallel to M, or
at least find a way to show this. Secondly, even if we could show that the wall
did remain the same as far as the M_z variation, shown in figure 2.3, was concerned,
it is certainly not obvious that exchange and anisotropy can still be ignored
when ϕ is large. This means that we have no reason, as yet, to believe that the
limiting velocity given by equation 2.55 applies in practice, although the inequality
given by equation 2.54 is certainly true.

We shall see in the next chapter that very good predictions of experimental
behaviour can be made using the very simple Landau–Lifshitz model described
here, provided we ensure that the angle ϕ does remain small. It is clearly more
important to move on to this work than to attempt to complicate the theoretical
treatment so that it fits some given experimental data, because the test of a good
theoretical model is what new experiments it can propose which will, perhaps,
completely contradict it and make us move on to something better. Bearing this
in mind, this chapter concludes with a brief look at some of the attempts which
have been made to include more terms in the Landau–Lifshitz equation and so
answer some of the questions raised here.

The most direct way of finding out what the exchange and anisotropy fields
really do in the moving wall is to write down the complete Landau–Lifshitz
equation in spherical polar coordinates, include the exchange and anisotropy
fields and make the simplifying assumption that the sample is infinite to avoid
the difficult magnetostatic field. The result of doing this is to obtain a set of
non-linear partial differential equations for the components of M. Unpublished
work by Walker, briefly outlined by Dillon (1963) took these lines and was
reported in greater detail by Schryer and Walker (1974) when the equations
were treated numerically. The results have some relevance to the bubble domain

materials so we shall consider them in chapter 3. From the point of view of a better understanding of what is happening in the moving wall, this approach is very unhelpful and there is a fundamental reason why. This is that the spherical polar coordinates, which obviously suit a problem involving a vector M which is assumed to have constant magnitude, have an axis of symmetry. This axis is obviously chosen to coincide with the easy axis of the magnetic material, z, so that we write

$$M_x = M_s \sin \theta \cos \phi$$

$$M_y = M_s \sin \theta \sin \phi \qquad (2.89)$$

$$M_z = M_s \cos \theta$$

corresponding to the coordinate system shown in figure 2.7. This was the co-ordinate system used in section 2.10.

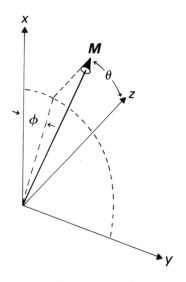

Figure 2.7 Normal spherical polar coordinates. z **is the polar axis**

The dynamic problem, however, involves motion along y and rotation of M about y. It follows that y is the axis of symmetry of the dynamic problem and the equations take on a very complicated form when a coordinate system using z as its axis of symmetry is employed.

This point was realised by Enz (1964) who used the coordinate system shown in figure 2.8. In these coordinates we have

$$M_x = M_s \cos \phi \sin \theta$$

$$M_y = M_s \sin \phi \quad\quad\quad (2.90)$$

$$M_z = M_s \cos \phi \cos \theta$$

which are the components of M in spherical polar coordinates when y is the polar axis. The angle ϕ is the complement of the angle which M makes with the polar axis.

Enz (1964) then took the undamped Landau–Lifshitz equation

$$\frac{dM}{dt} = -|\gamma|(M \times F) \quad\quad\quad (2.91)$$

and restricted the analysis to the case where ϕ is small so that $\cos \phi \approx 1$ and $\sin \phi \approx \phi$. Under these conditions, referring to figure 2.8 and equation 2.90 the components of dM/dt are

$$\frac{dM_x}{dt} = (M_s \cos \theta) \frac{\partial \theta}{\partial t}$$

$$\frac{dM_y}{dt} = (M_s) \frac{\partial \phi}{\partial t} \quad\quad\quad (2.92)$$

$$\frac{dM_z}{dt} = -(M_s \sin \theta) \frac{\partial \theta}{\partial t}$$

For the total field, F, we have the same anisotropy field as in previous work, equation 2.5, and an exchange field which involves only the derivatives with respect to y, because we assume that the wall remains plane. The magnetostatic field then involves only $\mu_0 M_x$ and $\mu_0 M_z$. In practical terms, we are considering the kind of situation illustrated in figure 2.2 where the wall behaves as a thin sheet in (x, z) and the divergence of M on the sample surface is considered to be negligible. From this point of view the components of F, from equations 2.5, 2.15 with 2.90, and 2.1, are

$$F_x = -\frac{2K_u}{M_s} \sin \theta + \frac{2A}{M_s} \left[\cos \theta \left(\frac{\partial^2 \theta}{\partial y^2} \right) - \sin \theta \left(\frac{\partial \theta}{\partial y} \right)^2 \right] + \mu_0 M_s \sin \theta$$
$$(2.93)$$

$$F_y = -\frac{2K_u}{M_s} \phi + \frac{2A}{M_s} \frac{\partial^2 \phi}{\partial y^2} \quad\quad\quad (2.94)$$

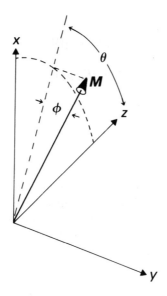

Figure 2.8 The coordinates used by Enz (1964). The polar axis is now the *y*-coordinate

and

$$F_z = -\frac{2A}{M_s}\left[\sin\theta\left(\frac{\partial^2\theta}{\partial y^2}\right) + \cos\theta\left(\frac{\partial y}{\partial\theta}\right)^2\right] + \mu_0 M_s \cos\theta \quad (2.95)$$

No applied field term has been introduced because we intend to study these equations from the point of view of the wall itself.

We now substitute equations 2.92 to 2.95 into equation 2.91 and obtain

$$\frac{-1}{\omega_s}\frac{\partial\theta}{\partial t} = -\varDelta^2 Q\left[\phi\left(\frac{\partial\theta}{\partial y}\right)^2 + \frac{\partial^2\phi}{\partial y^2}\right] + \phi\left(1 + Q\cos^2\theta\right) \quad (2.96)$$

by subtracting the *z*-component of equation 2.91, multiplied by $\sin\theta$, from the *x*-component, multiplied by $\cos\theta$. The *y*-component of equation 2.91 yields

$$\frac{-1}{\omega_s Q}\frac{\partial\phi}{\partial t} = \varDelta^2\frac{\partial^2\theta}{\partial y^2} - \cos\theta\,\sin\theta \quad (2.97)$$

Equations 2.96 and 2.97 have made use of the material parameters ω_s, Q and

Δ defined by

$$\omega_s = \mu_0 |\gamma| M_s \qquad (2.98)$$

$$Q = 2K_u/\mu_0 M_s^2 \qquad (2.99)$$

and

$$\Delta^2 = A/K_u \qquad (2.100)$$

to simplify the algebra.

Enz (1964) obtained a solution to equations 2.96 and 2.97, for the case in which $Q \ll 1$, which showed that the wall apparently contracted as the velocity increased. It is more interesting to attempt an approximate solution for finite Q, because when $Q \ll 1$ the wall width is very large anyway.

We can see what is happening from equation 2.97. In the static case, $\partial\phi/\partial t = 0$, we have

$$\Delta^2 \frac{\partial^2 \theta}{\partial y^2} - \cos\theta \sin\theta = 0 \qquad (2.101)$$

which was the equation 2.23, solved in section 2.6 for the static structure of the Landau–Lifshitz wall. When we went on to consider a moving wall, in section 2.10, we found that the angle ϕ had to increase from zero before any motion could occur. Let us suppose that ϕ has a maximum value $\hat{\phi}$ at the centre of the moving wall. The value of ϕ within the domains is clearly zero. It follows that $\partial\phi/\partial y$ must have a positive maximum, $\approx \hat{\phi}/\Delta$, on the $y < 0$ side of the wall and a negative minimum, $\approx -\hat{\phi}/\Delta$, on the $y > 0$ side. In form, $\partial\phi/\partial y$ is very much like the function $\cos\theta \sin\theta$ which occurs in equation 2.97. We may write

$$\frac{\partial\phi}{\partial y} \approx \frac{2\hat{\phi}}{\Delta} \cos\theta \sin\theta \qquad (2.102)$$

We are now considering a domain wall moving at constant velocity, however, and may use the relationship

$$\frac{\partial\phi}{\partial t} = \frac{d\phi}{dt} - \frac{\partial\phi}{\partial y} \frac{dy}{dt} \qquad (2.103)$$

to write equation 2.97 as

$$\Delta^2 \frac{\partial^2 \theta}{\partial y^2} - \left(1 + \frac{2v_y \hat{\phi}}{\omega_s Q \Delta}\right) \cos\theta \sin\theta = 0 \qquad (2.104)$$

because, for constant velocity, $d\phi/dt = 0$ and $dy/dt = v_y$.

The angle $\hat{\phi}$, discussed above, is related to the wall velocity by equation 2.79 and for zero damping, the case being considered here, equation 2.79 is simply

$$\hat{\phi} = v_y/2v_W \qquad (2.105)$$

when we substitute equation 2.55 for the Walker velocity

$$v_W = |\gamma|\mu_0 M_s \Delta/2 \qquad (2.106)$$

Equations 2.98, 2.105 and 2.106 may then be used to write equation 2.104 as

$$\left(\frac{\Delta^2}{1 + \dfrac{v_y^2}{2Qv_W^2}}\right) \frac{\partial^2\theta}{\partial y^2} - \cos\theta \sin\theta = 0 \qquad (2.107)$$

When equation 2.107 is compared with equation 2.101, the equation for the static wall structure, it is clear that we shall have the same solution for the structure of moving wall as we had for the stationary wall but that the wall width parameter will be reduced from Δ to

$$\Delta' = \Delta/\left(1 + \frac{v_y^2}{2Qv_W^2}\right)^{1/2} \qquad (2.108)$$

Equation 2.108 shows that changes in the wall structure would not be expected until the velocity approached $\sqrt{(2Q)}v_W$. In materials which have $Q > 1$, like the bubble domain materials we shall consider in chapters 3 and 4, this is a trivial result because the Landau–Lifshitz model only applies to velocities well below v_W anyway. For materials like YIG and Permalloy, however, Q is well below unity and the velocity $\sqrt{(2Q)}v_W$ should be considered the limiting velocity from the point of view of the Landau–Lifshitz model, not the higher velocity, v_W.

The significance of the velocity $\sqrt{(2Q)}v_W$ is easier to see when we write

$$v_c = \sqrt{(2Q)}v_W \equiv |\gamma|\sqrt{(\mu_0 A)} \qquad (2.109)$$

which follows from a combination of equations 2.99, 2.100 and 2.106. This agrees with the critical velocity deduced by Enz (1964) apart from a numerical factor close to unity. When the critical velocity is written in this way, $|\gamma|\sqrt{(\mu_0 A)}$, it is clear that its value is rather material independent because the exchange constant, A, is close to 10^{-11} J/m in nearly all materials and $|\gamma| = 1.76 \times 10^{11}$ (rad/s) per T applies to most magnetic materials of practical interest for use at normal temperatures. When these values of A and $|\gamma|$ are substituted into equation 2.109 a critical velocity of 624 m/s is obtained. This will be discussed further at the end of this chapter.

The idea that a domain wall might contract in width as its velocity increases, as shown by equation 2.108, is also found in the work of Palmer and Willoughby (1967), Feldtkeller (1968), Schlomänn (1971a, 1971b), Bourne and Bartran (1972), Bartran and Bourne (1972, 1973) and Schryer and Walker (1974). Most of this work was directed towards the problem of domain wall motion in very thin Permalloy films of the kind which will be considered in chapter 5. Schlömann (1972) extended his work to show that the contraction in the wall width might explain the saturation velocity which is observed in some wall motion experiments in very thin films.

2.18 Conclusions

In this chapter we have looked at the static structure and then the dynamic properties of the Landau–Lifshitz domain wall. This is a particularly simple kind of domain wall, an idealisation, which should approximate to the true wall structure found in samples of practical dimensions provided the easy direction of magnetisation lies in the plane of the sample and the thickness of the sample is very much greater than the wall width parameter, Δ, given by equation 2.29.

The main dynamic results obtained in this chapter show that the wall should take up a velocity which depends linearly upon the drive field, equations 2.52 and 2.53, and appear to have a constant mass, given by equation 2.77. Both these results depend upon the drive field being very small, less than the value $\alpha\mu_0 M_s/2$, equations 2.54 or 2.72, associated with the Walker velocity, equations 2.55 or 2.106. On top of this restriction we find what may be an even stronger restriction upon the wall velocity in the form of the critical velocity, $v_c = |\gamma|\sqrt{(\mu_0 A)}$, derived in section 2.17 as equation 2.109. This critical velocity is certainly the more important one in YIG, as shown in table 2.1, and this point was made by Safiullah and Teale (1978) in the most recent work concerning the low drive wall mobility in YIG.

There is a further restriction which must be made before the simple Landau–Lifshitz theory of wall dynamics may be applied to a practical problem. This concerns our assumption that the wall remains plane as it moves forward. The problem of the internal structure changing because of wall motion was considered in section 2.17 but that was still done within the framework of the one-dimensional representation given by equations 2.30 and 2.31. The question to be considered now is that of possible wall bending.

The situation which could arise is shown in figure 2.9 where a domain wall is shown moving to the right, as in previous examples, but is now subject to some additional loss or drag at the surface of the sample which causes the moving wall to become bowed in the way shown. A very similar kind of wall bowing will be discussed in chapter 5 in connection with wall motion in conducting media and, there, the wall bowing will be found to be in the opposite sense to that shown in

Table 2.1 Some measured and calculated parameters for nickel ferrite at 201 K (Galt, 1954) and for YIG at 295 K (Harper and Teale, 1969)

	Nickel ferrite	YIG	Equations
$\mu_0 M_s$ (T)	0.42	0.17	measured
A (J/M)	1.09×10^{-11}	4.3×10^{-12}	measured
K_u (J/m^3)	5400	580	measured
$\|\gamma\|$ (rad/s per T)	1.9×10^{11}	1.76×10^{11}	measured
α	4×10^{-3}	10^{-4}	measured
$\Delta = (A/K_u)^{1/2}$	0.45×10^{-7}	0.86×10^{-7}	2.29
$Q = 2K_u/\mu_0 M_s^2$	7.7×10^{-2}	5.0×10^{-2}	2.60, 2.99
$l = 2Q\Delta$	7×10^{-9}	8.6×10^{-9}	2.117
$B_{max} = \alpha\mu_0 M_s/2$	8.4×10^{-4}	0.085×10^{-4}	2.54, 2.72
$v_W = \|\gamma\|\mu_0 M_s\Delta/2$	1800	1290	2.55, 2.106
$v_c = \|\gamma\|\sqrt{(\mu_0 A)}$	700	410	2.109

figure 2.9 because eddy-currents effectively shield the moving wall from the drive field at the centre of the sample whereas the full force of the drive field will be felt on the sample surface.

The possibility of additional damping due to surface roughness, which might cause the kind of wall bowing shown in figure 2.9, has been considered from the point of view of domain wall motion by Huang (1969) in an important paper. Here, we shall only consider the conditions under which wall bowing might occur and thus invalidate the simple model which has been built up so far. This is done very easily if a polar coordinate system is used, as shown in figure 2.9, with its origin chosen as the centre of wall curvature; point A in figure 2.9.

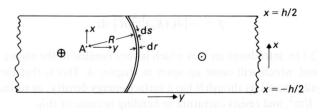

Figure 2.9 The wall region may become bowed as it moves forward, taking on a radius of curvature, R

The z-component of the exchange field in this coordinate system is

$$(B_{ex})_z \;=\; \frac{2A}{M_s^2} \frac{1}{r} \frac{\partial}{\partial r}\left(r\,\frac{\partial M_z}{\partial r}\right) \tag{2.110}$$

which follows from equation 2.15 because it may be assumed that M varies only with r in the local coordinates of figure 2.9. This means that we have extended equation 2.30 to represent the wall structure as

$$M_z \;=\; M_s \tanh\,(R-r)/\varDelta \tag{2.111}$$

Expanding equation 2.110 gives two terms

$$(B_{ex})_z \;=\; \frac{2A}{M_s^2}\left(\frac{1}{R}\frac{\partial M_z}{\partial r} + \frac{\partial^2 M_z}{\partial r^2}\right) \tag{2.112}$$

when we set $r = R$ in order to find the value of the field at the wall itself. Substitution of equation 2.111 into equation 2.112 leaves only the first term at $r = R$ and this is

$$(B_{ex})_z \;=\; -(2A/M_s R\,\varDelta) \tag{2.113}$$

Equation 2.113 shows that there will be a restoring force upon the wall if it becomes bowed which is given by $2M_s(B_{ex})_z\,ds$, acting upon the wall element ds in figure 2.9 and tending to make the wall plane once more. From equation 2.113 the expression for this force is

$$f \;=\; -(4A/R\varDelta)ds \tag{2.114}$$

and when the wall width parameter

$$\varDelta \;=\; (A/K_u)^{1/2} \tag{2.115}$$

is introduced, equation 2.114 may be written as

$$f \;=\; -[4(AK_u)^{1/2}/R]\,ds \tag{2.116}$$

Equation 2.116 introduces an idea which is very familiar in the statics of bubble domains, and which will come up again in chapter 4. This is that the Landau–Lifshitz wall behaves as though it has a surface energy density, or surface tension, $4(AK_u)^{1/2}$ J/m^2, and resists curvature or bending because of this.

The problem is made particularly simple if the material characteristic length

$$l \;=\; 4(AK_u)^{1/2}/\mu_0 M_s^2 \tag{2.117}$$

is introduced. We may then write equation 2.113 as

$$(B_{ex})_z = -\mu_0 M_s l/2R \qquad (2.118)$$

to give the strength of the restoring force due to the wall curvature, R, or bowing, relative to $\mu_0 M_s$. Clearly, in a material in which $l \ll h$, where h is the thickness of the sample as shown in figure 2.9, a drive field well below $\mu_0 M_s$ may cause significant wall bowing; that is, $R \approx h$. The problem may be summarised by saying that the drive field must not be allowed to exceed

$$\hat{B} = \mu_0 M_s l/h \qquad (2.119)$$

in magnitude if the assumption that the wall remains plane is to be expected to hold.

From the point of view of the two materials discussed in this chapter, nickel ferrite and yttrium iron garnet, table 2.1 gives an overall numerical view. The characteristic length, l, of these materials is very small so that if the field given by equation 2.119 is to be well above the maximum drive field used in the experiments, very thin samples must be used. In the case of nickel ferrite, Galt (1954) used drive fields up to 30 μT so that, from equation 2.119 and the figures given in table 2.1, the sample thickness should have been kept below 100 μm. Similarly, samples below 150 μm should have been used by Harper and Teale (1969) whose drive fields were close to 10 μT. Dimensional considerations of this kind are very important when experimental data on very low Q or very low anisotropy material are being considered. The point will come up again in chapter 5.

Another restriction which has been made in this chapter is that only materials with uniaxial anisotropy have been considered. At first sight this might appear to be a very serious restriction because both ferrites and garnets are often termed cubic materials. This, of course, is only true when the ferromagnetism is ignored: there can be no cubic magnetic crystal point groups which show ferromagnetism. This has been well explained by Ascher (1966).

Both nickel ferrite and YIG belong to the uniaxial magnetic point group $\bar{3}\underline{m}$ and in the case under consideration here, that of a 180° domain wall, the argument would be that the thin wall region is dominated by the $\bar{3}\underline{m}$ symmetry of the domains upon either side and the use of a simple uniaxial model, equation 2.3, is sensible. If the wall is not a 180° one, but involves a change in the direction of M through some other angle, the wall should really be considered a magnetic crystal boundary, that is, a grain boundary in a magnetic polycrystal. Lilley (1950) has considered the statics of a variety of domain walls involving complications of this kind and has also brought in magnetostriction. This work by Lilley was before the developments in magnetic crystallography, which were mentioned above (Ascher, 1966), that really began with the important work *Colored Symmetry* by Shubnikov and Belov (1964). Later references on magnetic crystals are Opechowski and Guccione (1965), Birss (1966) and O'Dell (1970).

The dynamics of walls which separate domains of different magnetic crystal orientations, that is other than simple 180° domain walls, should really be considered as the dynamics of phase boundaries. This point of view has been taken by Kikuchi (1960) and by Cahn and Kikuchi (1966) in some very general work which would also apply to the dynamics of domain walls with the structure described by Bulaevskii and Ginzburg (1964, 1970), discussed in section 2.7.

Finally the problem of magnetostriction must be considered. This will certainly produce a resistance to wall motion because, even if there is no change in the strain within the domains upon either side of a 180° domain wall, the wall region itself may be strained because M is rotating away from the easy direction of magnetisation inside the wall. The part played by magnetostriction in ferromagnetodynamics has not been considered very much in the literature. Callen (1958) discusses the problem from a general point of view and Uchiyama *et al.* (1976) have considered the interaction between a moving domain wall and an acoustic wave, an interaction which may explain the interesting results of Tsang and White (1974) and Tsang *et al.* (1975, 1978) which were discussed in section 1.4.

3 Straight Wall Motion in Magnetic Bubble Films

3.1 Introduction

The Landau–Lifshitz model for magnetic domain wall motion was established in chapter 2 and was found to apply only under very special conditions. These conditions were that the sample being considered should have its easy direction of magnetisation lying in-plane, that its thickness should be much greater than the domain wall width and, finally, that the field which is applied to drive the wall should be well below the critical threshold, $\alpha\mu_0 M_s/2$, given by equation 2.72.

In this chapter we are going to look at some experimental data which throws considerable light upon the Landau–Lifshitz model and leads us to look at domain wall motion from a very different point of view. This work has been done using thin single crystal layers of magnetic garnets of the kind which are prepared for magnetic bubble domain devices (O'Dell, 1974) and the important point about these materials is that they do not satisfy the conditions needed for the Landau–Lifshitz model which were listed above.

3.2 Magnetic Bubble Domain Materials

Experimental work in ferromagnetodynamics has always been constrained very severely by the availability of suitable materials and the way in which these may be prepared. The magnetic materials which are, at present, developed to the highest degree of crystalline perfection and with the widest range of compositions are the epitaxial garnet films which are used in magnetic bubble domain devices.

These magnetic bubble films do not appear to meet the special conditions needed for a test of the Landau–Lifshitz model. In fact, they seem to be opposed to these conditions in two ways. Firstly, the easy direction of magnetisation is normal to the plane of the film so that we should not expect the simple Landau–Lifshitz wall structure, derived in section 2.6, to apply in the static case, let alone the dynamic. Secondly, even if we had such a bubble film which had an in-plane easy direction, or if we could expect the Landau–Lifshitz wall structure to apply for some special reason, these magnetic bubble films tend to have very low values of magnetisation. A typical value of $\mu_0 M_s$ might be 200×10^{-4} T so that the critical field, $\alpha\mu_0 M_s/2$, given by equation 2.72, is going to be 0.01×10^{-4} T, if the value of α is around 10^{-4}, as we would expect in a low loss garnet. This means we shall need to keep the applied field below 0.001×10^{-4} T if we are to observe wall motion in bubble films which follows the Landau–Lifshitz model.

It will be impossible to work with crystals having the slightest imperfections.

For the experimentalist in ferromagnetodynamics, however, the attractions of the magnetic bubble films are so considerable that some way must be found to overcome the difficulties described above. Bubble films are so attractive because the out-of-plane magnetisation means that optical techniques may be used to observe the dynamic process, a wide range of compositions are available and, on top of these important experimental advantages, there is the technical interest in the ferromagnetodynamics of these materials. Let us see if it is possible to propose an experiment which uses one of these convenient materials and also adds to our understanding of the Landau–Lifshitz model by illustrating some of its predictions.

3.3 The Domain Wall Structure in a Magnetic Bubble Film

Let us first look at the domain wall structure which is expected in a magnetic bubble film to see just how different it is to the simple Landau–Lifshitz structure which was considered in chapter 2.

The problem is illustrated in figure 3.1 where a cross-section of the wall region is shown. The wall separates two domains which are uniformly magnetised, in opposite directions, along the easy direction of magnetisation which lies perpendicular to the plane of the film. We have drawn the wall thickness small in comparison with the film thickness, h, and have tentatively shown the magnetisation lying in-plane at the centre of the wall. This kind of wall was first discussed in connection with figure 1.4.

There is a very strong magnetic field circulating around a wall of this kind, quite regardless of the internal structure of the wall. This is shown in figure 3.2. From the point of view of the wall structure, the most important part of this field is the strong in-plane component, of the order of $\mu_0 M_s$, which exists at the top and the bottom of the wall, as shown in figure 3.2. The effect of this field is to make the magnetisation turn along the positive y-direction at the top of

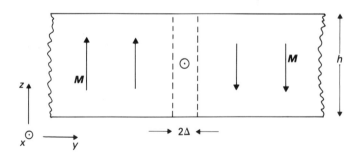

Figure 3.1 A domain wall separating two domains in a magnetic bubble film. At the exact centre of the wall the magnetisation lies in the plane of the film. This wall has a clockwise screw-sense at the centre

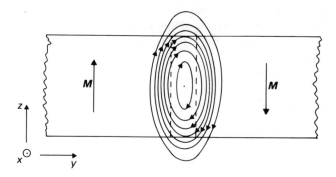

Figure 3.2 Regardless of the internal structure of the wall a strong magnetic field must circulate around the wall in a magnetic bubble film

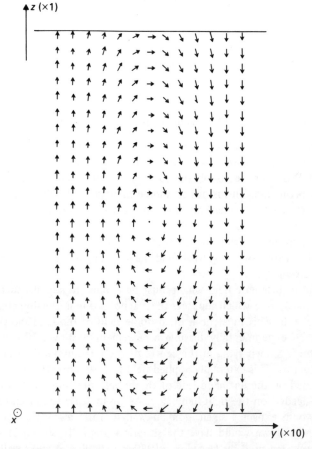

Figure 3.3 The detailed static wall structure for a magnetic bubble film found by Della Torre *et al.* (1975)

the wall region and along the negative y-direction at the bottom. An exact calculation of the detailed wall structure is not easy but it has been successfully completed by Della Torre *et al.* (1975) and their solution is shown in figure 3.3.

This solution applies when the wall structure is assumed to be independent of x, a condition which is not easy to obtain in practice, and it is obvious from figure 3.3 that it is only at the extract centre of the film that the wall structure is similar to the simple Landau–Lifshitz solution of section 2.6. At the centre of the film the magnetisation does rotate about the y-axis as we pass from one domain to the other and this rotation may take either sign, as with the Landau–Lifshitz wall. A clockwise rotation was chosen for figure 3.1.

At the top and the bottom of the wall region, figure 3.3 shows that the wall structure is similar to the Néel wall discussed in section 2.7. The screw-sense of these Néel sections is not arbitrary. Looking along the positive x-axis, M must rotate anticlockwise at the top of the wall, as we move along y, and clockwise at the bottom.

The magnetic bubble film thus has a two-dimensional wall structure which depends upon y and z and is quite different to the simple one-dimensional structure, depending only upon y, which was the subject of chapter 2.

3.4 The Stability of Straight Walls in Magnetic Bubble Films

In the picture frame samples, discussed in chapters 1 and 2, which were cut from single crystals in such a way as to have the easy direction of magnetisation lying in-plane, it was possible to obtain an isolated and straight domain wall for dynamic experiments. This is not so simple in magnetic bubble films because of the strong external field which was shown in figure 3.2. If an isolated and straight wall did exist momentarily it would be completely unstable and would ripple out into a long meandering wall so that the whole film became effectively demagnetised. Clearly, some external field must be applied if we require a straight and isolated wall.

The kind of magnetic field which must be applied to support such a straight wall is obviously one which supports the magnetisation in the domains on either side of the wall. This must be a quadrupole field of the kind shown in figure 3.4 where an experimentally feasible arrangement is shown. The thin bubble domain film is shown lying in the centre of the quadrupole field produced by eight conductors. This externally applied field now supports the upward magnetisation needed on the left-hand side of the bubble film and the downward magnetisation needed on the right-hand side. As we increase the magnitude of this field we would expect a single domain wall to form along the centre of the bubble film and we would have the situation originally shown in figure 3.1. This idea was first used by Hagedorn (1970) to stabilise straight walls in orthoferrite crystals, which were the first materials used for magnetic bubble work.

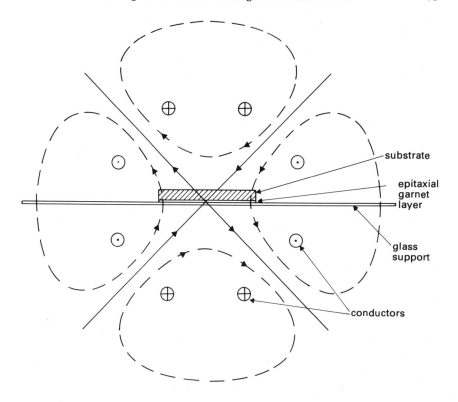

Figure 3.4 Showing a magnetic bubble film supported at the centre of a quadrupole field produced by eight conductors carrying equal currents of the polarities indicated. This field will stabilise a straight isolated wall in the film

Schlömann (1974) gives an expression for the value of $\partial B_z/\partial y$ needed to stabilise a straight wall, which depends upon the material parameters and a very large field gradient may be needed. For bubble films which have a value of

$$Q = 2K_u/\mu_0 M_s^2 \tag{3.1}$$

of about 5, which is typical for normal garnet films, the gradient $\partial B_z/\partial y$, will be of the order of $\mu_0 M_s/h$. If the Q value can be increased a smaller gradient field can be used. It follows that the experimental conditions are fairly easy to realise for orthoferrites because the Q is very high, and the thickness, h, of the crystals is typically 100 μm. The much thinner garnet bubble films require very large stabilising gradients.

3.5 Producing a Landau–Lifshitz Wall Structure in a Magnetic Bubble Film

We have gone into some experimental detail here because it is important to realise that the stabilising field needed for a straight wall in a bubble film has a large intensity and also has the shape shown in figure 3.4. It is, of course, not possible to produce a magnetic field in which $\partial B_z/\partial y$ has a value and $B_y = 0$, that is a pure gradient field, by using conductors or permanent magnets which are placed outside the garnet film. This follows because we must have curl $B = 0$ in the free space in which we wish to place our garnet sample. The relationship

$$\partial B_z/\partial y = \partial B_y/\partial z \qquad (3.2)$$

must be satisfied regardless of the way in which we produce our stabilising gradient, $\partial B_z/\partial y$. The field which satisfies equation 3.2 is the quadrupole field shown in figure 3.4.

If we compare the magnetic field shown in figure 3.2, the field associated with the wall region, with the field shown in figure 3.4, the quadrupole field, it is immediately obvious that a superposition of these two fields could produce a pure B_z field which reverses sign as we pass through the plane $y = 0$. This would happen when the B_y field of figure 3.4 cancelled out the B_y field of figure 3.2, these being in the opposite direction whereas the B_z fields in the two diagrams are in the same direction. This condition gives a qualitative reason why the field shown in figure 3.4 can support a stable straight domain wall in a bubble film. Once the B_y field of the wall is cancelled out, the boundary conditions for the wall become almost identical to the boundary conditions for an infinite sample: the problem dealt with in section 2.6. The applied gradient field not only stabilises a straight wall but also produces a Landau–Lifshitz wall structure in the magnetic bubble film. There is no longer any B_y field in the wall region which will cause the magnetisation there to take on the y-component shown in figure 3.3. The same argument shows us why, when we neglect the effect of anisotropy, the field gradient required is of the order of $\mu_0 M_s/h$. It is because the component B_y of the field shown in figure 3.2 is of the order of $\mu_0 M_s$ at the top and bottom surfaces of the bubble film (O'Dell, 1978a).

It would thus appear that we have a method of producing isolated straight domain walls in magnetic bubble films which can have the simple one-dimensional Landau–Lifshitz structure covered by the theory given in chapter 2. There is, however, one very difficult experimental problem to overcome and this is that the garnet film, shown in figure 3.4, must be located in the quadrupole field so that its centre plane lies exactly in the $z = 0$ plane of the field. Now this might be possible if the quadrupole field were generated by eight very accurately fixed conductors, as shown in figure 3.4. In practice, however, the quadrupole field must be generated by a permanent magnet arrangement and the exact location of its plane of symmetry is uncertain.

If the garnet film is placed off the plane of symmetry we effectively have a superposition of the quadrupole field and a uniform B_y field. The intensity of this in-plane field increases linearly with the distance which we move away from the plane $z = 0$. It follows that the experiment we are concerned with is one in which we have produced a Landau–Lifshitz wall but have a uniform in-plane field applied, the value of which is not accurately known until we can find from the results of the experiment exactly where the $z = 0$ plane lies. This could be done by lifting the sample up and down on the support shown in figure 3.4.

It was just this difficulty which precipitated a series of very interesting experiments by de Leeuw and his co-workers, beginning with de Leeuw (1973). Because the dynamic behaviour of the special kind of domain wall we have described here does depend upon the magnitude of the in-plane field which is applied, as well as depending upon the magnitude of the drive field, it is possible to locate the $z = 0$ plane and know the experimental conditions. Our first problem is to understand the effect of the in-plane field and we shall then look at the results of these experiments.

3.6 Motion of a Landau–Lifshitz Wall when a Magnetic Field is Applied Perpendicular to the Wall Plane

In chapter 2, figure 2.5, we found that wall motion, in the Landau–Lifshitz model, implies that the magnetisation must be deflected out of the plane of the wall by an angle ϕ and precess around the y-axis, which is the direction of motion. The part of the Landau–Lifshitz equation which really shows the origin of this kind of wall motion is the z-component, which was written as

$$\frac{-1}{|\gamma|} \frac{dM_z}{dt} = -\mu_0 M_s^2 \cos\phi \sin\phi - \lambda B_z \tag{3.3}$$

which was equation 2.46. Here we see the out of plane component of M, $M_s \sin\phi$, and the magnetic field at the centre of the wall, $\mu_0 M_s \cos\phi$, giving the major contribution to the motion, $\partial M_z/\partial t$, and having maximum effect when $\phi = \pi/4$ where we have the Walker limiting velocity

$$v_W = |\gamma|\mu_0 M_s \Delta/2 \tag{3.4}$$

previously given as equation 2.55.

Now the z-component of $(M \times B)$ is $(M_x B_y - M_y B_x)$ and only the second term has come into our argument above because B_y was zero in the work covered in chapter 2. It seems sensible to suggest that the velocity of the wall might be increased if we could apply a field B_y in such a way that the term $M_x B_y$ adds to the term $-M_y B_x$ which is already involved.

To study this problem we must first establish a relationship between an externally applied B_y field and the field which appears within the wall. We again consider an infinite medium and this simplifies the problem at once because we know that the magnetisation within the wall itself cannot produce any magnetic field in the y-direction, this direction being normal to the plane of the wall. It follows that we can remove the wall region and calculate the magnetic field which is produced in the slit, width 2Δ, which now separates the two domains.

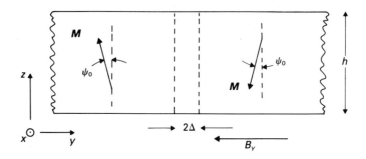

Figure 3.5 A bubble film, thickness h, is placed in a large in-plane field B_y. The magnetisation within the domains tilts by an angle ψ_0. A negative B_y is chosen because this supports high velocity motion for the positive screw-sense walls which have been shown in all the previous diagrams. The internal wall structure is not shown in this diagram; it depends upon the motion and the applied fields

The first step in this calculation is to find out what is happening in the domains as a result of the applied field, B_y. We would expect to find that the magnetisation is tilted away from the easy direction, as shown in figure 3.5, by an angle, ψ_0, and to find ψ_0 we simply solve the equation

$$\mathbf{M} \times \mathbf{F} = 0 \tag{3.5}$$

introduced as equation 2.2 in chapter 2. There is no exchange contribution to $\mathbf{F}$, because we assume that ψ_0 is a constant in each domain. We have only to consider the anisotropy field, magnetostatic field and the applied field, B_y. The components of $\mathbf{F}$ are then given by

$$F_x = 0$$

$$F_y = \mu_0 M_s \sin \psi_0 - \mu_0 M_s Q \sin \psi_0 + B_y \tag{3.6}$$

$$F_z = \mu_0 M_s \cos \psi_0$$

and the only component of equation 3.5 which we need consider is

$$(M \times F)_x = \mu_0 M_s^2 Q \cos \psi_0 \sin \psi_0 - B_y M_s \cos \psi_0 \qquad (3.7)$$

Equation 3.7 vanishes when

$$\sin \psi_0 = B_y / \mu_0 M_s Q \qquad (3.8)$$

The angle ψ_0 is shown negative in figure 3.5 because we have chosen a negative value of B_y in this diagram to tie up with the dynamic discussion which will follow. To conclude the static calculation first, if we treat the wall region as a narrow slit between the two domains, then F_y is simply continuous across the slit, excluding, of course, the anisotropy field which is an effective field representing a material property. Equations 3.6 and 3.8 thus give

$$F_y = B_y / Q + B_y \qquad (3.9)$$

the sum of the field due to the tilted magnetisation within the domains and the applied field.

We now return to our dynamic problem and make the drastic assumption that the moving wall has the Landau–Lifshitz form. We shall be able to show that this assumption is valid for the moving wall whereas it is quite obvious that we could not expect a Landau–Lifshitz structure under the static conditions shown in figure 3.5. It turns out that the static structure of the wall is very much like the Néel structure referred to in section 2.7 (O'Dell, 1978a).

If we assume the Landau–Lifshitz structure it means that we must have ψ_0 small and that the magnetisation rotates about the y-axis as we pass through the wall region, making a similarly small angle, ϕ, with respect to (x, z). Following the same argument as section 2.10, we take the Landau–Lifshitz equation

$$\frac{-1}{|\gamma|} \frac{dM}{dt} = (M \times F) - \lambda[F - (F \cdot M)M/M_s^2] \qquad (3.10)$$

and consider the centre point of the moving wall where F_y is given by equation 3.9 and

$$F_x = \mu_0 M_s \cos \phi$$
$$\qquad (3.11)$$
$$F_z = B_z$$

We consider the wall moving at a constant velocity, v_y, as a result of the drive field B_z. The x- and y-components of equation 3.10 must both vanish, as equations 2.44 and 2.45 did in our previous work. We find that, setting the

x-component equal to zero gives

$$B_z M_s \sin \phi - \lambda \{ \mu_0 M_s \cos \phi \sin^2 \phi \; - [B_y(Q+1)/Q] \cos \phi \sin \phi \} = 0 \qquad (3.12)$$

while the y-component gives

$$-B_z M_s \cos \phi + \lambda \{ \mu_0 M_s \cos^2 \phi \sin \phi + [B_y(Q+1)/Q] \cos^2 \phi \} = 0 \qquad (3.13)$$

Equations 3.12 and 3.13 are both satisfied when

$$B_z = \mu_0 \lambda \{ \cos \phi \sin \phi - [B_y(Q+1)/\mu_0 M_s Q] \cos \phi \} \qquad (3.14)$$

We now consider the z-component of the Landau–Lifshitz equation, equation 3.10. M_z must change from $-M_s \cos \psi_0$ in the right-hand domain to $+M_s \cos \psi_0$ in the left-hand domain as the wall moves from left to right. Following the argument which led to equation 2.48, we now have

$$\frac{dM_z}{dt} = \frac{M_s v_y \cos \psi_0}{\Delta} \qquad (3.15)$$

and, using equation 3.8 to express $\cos \psi_0$ in terms of B_y, M_s and Q, we have the z-component of equation 3.10 as

$$\frac{-M_s v_y}{|\gamma| \Delta} [1 - (B_y/\mu_0 M_s Q)^2]^{1/2} = \qquad (3.16)$$

$$\{ [B_y M_s(Q+1)/Q] \cos \phi - \mu_0 M_s^2 \sin \phi \cos \phi \} - \lambda B_z$$

From equation 3.14 we can recognise at once that the first terms on the right-hand side of equation 3.16 are equal to $-B_z M_s^2/\lambda$ and can write

$$v_y = \frac{|\gamma| M_s \Delta (1 + \lambda^2/M_s^2)}{\lambda [1 - (B_y/\mu_0 M_s Q)^2]^{1/2}} B_z \qquad (3.17)$$

If we neglect the term $(B_y/\mu_0 M_s Q)^2$ compared to unity, and this will nearly always be possible in a practical case, equation 3.17 is exactly the same result for the velocity–field relationship as we obtained in chapter 2, as equation 2.51, and it would seem that the externally applied in-plane field, B_y, has had no effect upon the wall motion. What is quite different, however, is the range of B_z over which equation 3.17 applies. From equation 3.14 we can see that very large values of B_z can be applied if B_y is made negative and ϕ is kept near zero. This is a very important result because it is precisely when ϕ is near zero that we can accept our initial assumption that the moving wall has the Landau–Lifshitz structure.

The point here is that our in-plane field, B_y, is being used to counteract the dynamic forces which would normally cause M to take up an angle, ϕ, with respect to the plane of the wall. We saw in chapter 2 how the angle ϕ increased with velocity and could reach a maximum value of $\pi/4$, corresponding to the rather artificial Walker limiting velocity

$$v_W = |\gamma|\mu_0 M_s \Delta/2 \tag{3.18}$$

We now have a situation in which we could increase the drive field, B_z, to increase the velocity, and also increase the in-plane field, B_y, to keep the angle ϕ small. We know that the Landau–Lifshitz model is fairly accurate when ϕ is small so that we now have an experiment which should follow this model.

Returning to equation 3.14, we differentiate this with respect to ϕ and find the maximum value of B_z, for a given value of B_y, is given when $\phi = \hat{\phi}$ where

$$\hat{\phi} = \arcsin\left\{[-b + (b^2+8)^{1/2}]/4\right\} \tag{3.19}$$

in which

$$b = |B_y|(Q+1)/\mu_0 M_s Q \tag{3.20}$$

In equations 3.19 and 3.20, and in what follows, we have chosen B_y to be directed along the negative y-axis so that it supports a wall which moves at the maximum possible velocity with a positive value of $\hat{\phi}$. This is an important point which brings in the choice we have tacitly made concerning the wall screwsense and will be discussed in a later section. We can now substitute equation 3.19 into equation 3.14 and put the result into equation 3.16 neglecting terms in λ^2/M_s^2 compared to unity. Expressed in terms of the Walker velocity, equation 3.18, we now find that the peak or maximum velocity of the wall is given by

$$\hat{v}_y/v_W = 2(\cos\hat{\phi}\sin\hat{\phi} + b\cos\hat{\phi})/$$
$$\left\{1 - [b/(Q+1)]^2\right\}^{1/2} \tag{3.21}$$

and it is obvious that $\hat{v}_y$ can exceed v_W because b can exceed unity when the value of Q is large compared to unity.

The theoretical results of this section have been summarised in figures 3.6 and 3.7. Figure 3.6 shows how the maximum value of ϕ, given by equation 3.19,

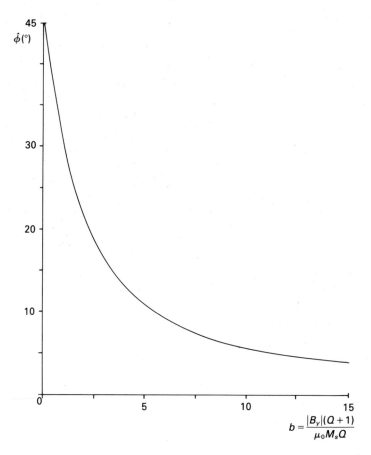

Figure 3.6 **The maximum angle which M can make with the wall plane, $\hat{\phi}$, given by equation 3.19, decreases as the in-plane field, B_y, is increased**

decreases from the value of $\pi/4$ as the in-plane field B_y is increased. The field B_y has been expressed by the parameter b and, because $Q \gg 1$ in the experiments we shall discuss, this parameter is effectively $B_y/\mu_0 M_s$.

Figure 3.7 shows the complete picture. Here, values of ϕ have been substituted into equation 3.14, for a range of values of b, and $B_z/\mu_0\lambda$ has been calculated. We can now see how the angle ϕ varies with drive field when this is reduced below the critical value at which $\phi = \hat{\phi}$. We see that, once b exceeds about 3.0, there is a very narrow range of values of $B_z/\mu_0\lambda$ over which the angle ϕ is close to zero. From the point of view of the experimentalist, we have moved right away from the origin of figure 3.7, where the drive field had to be kept below about $0.2\mu_0\lambda$ in order to keep ϕ below $10°$, to values of drive field around $6\mu_0\lambda$ illustrated by the curve marked $b = 6.0$. If we can experiment in this region, we shall be using drive fields which will be well above the intrinsic

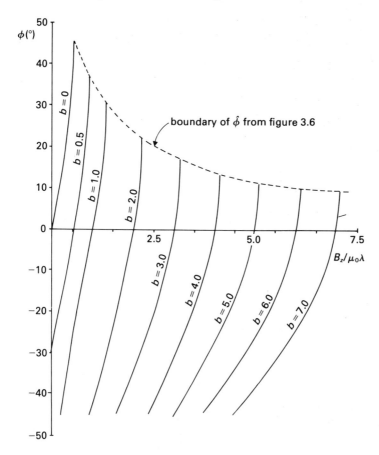

Figure 3.7 Equation 3.14 gives a set of curves for various values of in-plane field, B_y, defined by the parameter b, equation 3.20, which show how ϕ varies with drive field B_z. The curves terminate at $\hat{\phi}$ and begin, quite arbitrarily at $-\pi/4$ because the only region of interest is near $\phi = 0$

coercivity of the material, due to its imperfections, and should observe velocities well above the Walker limiting velocity, as shown by equation 3.21. Such experiments have been done and will be discussed in the next section.

3.7 Experimental Work on Wall Motion when a Field is Applied Perpendicular to the Wall Plane

The technique of producing isolated and stable straight domain walls in magnetic bubble films, which was described in section 3.4, was used by de Leeuw (1973) in the first of a series of most interesting experiments.

The epitaxial garnet film used by de Leeuw (1973) in his first experiments had a composition $Y_{2.9}La_{0.1}Fe_{3.8}Ga_{1.2}O_{12}$ and its material parameters are listed in table 3.1 under film No. 1. The film had a very low $\mu_0 M_s$ and a high Q so that a straight isolated domain wall could be stabilised with a quadrupole field of a permanent magnet arrangement. Pulsed drive fields, B_z, were applied to the wall and its velocity was determined photometrically. When an in-plane field was applied by moving the film above or below the field centre of the quad-rupole magnet, very high wall velocities were observed, up to ten times the Walker limiting velocity given in table 3.1, and in later experiments by Rijnierse and de Leeuw (1973) a clear maximum velocity was observed. If the drive field was increased above the value at which the maximum velocity was observed, the behaviour of the wall became quite different in that the velocity dropped dramatically and appeared to take on a fairly drive-independent value.

Later experiments by de Leeuw (1974) showed that the wall could begin to move at the maximum velocity when a drive field just above the critical value was applied but that this high velocity would only persist for a few tens of nanoseconds and then the low velocity would be observed. More data on the way in which the maximum observed velocity depended upon the value of the applied in-plane field were given by de Leeuw and Robertson (1975) using a different single crystal film: No. 2 in table 3.1. Data for a third sample, No. 3 in table 3.1, were given by de Leeuw (1977b) and de Leeuw *et al.* (1978).

Table 3.1 The material parameters for the magnetic bubble films used in the experiments of de Leeuw (1973) and Rijnierse and de Leeuw (1973), Film No. 1, of de Leeuw and Robertson (1975), Film No. 2, and of de Leeuw (1977b) and de Leeuw *et al.* (1978), Film No. 3. The values of Δ, Q and v_W have been calculated from equation 2.29, 3.1 and 3.4

	Film no. 1	Film no. 2	Film no. 3
$\mu_0 M_s$	68×10^{-4} T	60×10^{-4} T	46×10^{-4} T
A	10^{-12} J/m	10^{-12} J/m	10^{-12} J/m
K_u	550 J/m^3	310 J/m^3	212 J/m^3
$\vert\gamma\vert$	2.06×10^{11}	1.86×10^{11}	1.87×10^{11}
h	3.9×10^{-6} m	2.6×10^{-6} m	4.14×10^{-6} m
Δ	0.43×10^{-7} m	0.57×10^{-7} m	0.69×10^{-7} m
Q	30	22	25
v_W	30 m/s	32 m/s	30 m/s

All these data have been collected together in figure 3.8 where the experimentally observed velocities have been normalised to the respective Walker velocities for the various samples and the values of in-plane field expressed by means of the parameter b, defined by equation 3.20. In this way the theoretical result equation 3.21 may also be displayed. It should be noted that, despite the rather complicated appearance of equation 3.21, when both equations 3.19 and 3.20 are substituted to make $\hat{v}_y/v_W$ a function of b and Q only, it rapidly becomes asymptotic to the straight line

$$\hat{v}_y/v_W = 2|B_y|(Q+1)/\mu_0 M_s Q \qquad (3.22)$$

when the term $[b/(Q+1)]^2$ is neglected in comparison to unity, an assumption which is always valid in any practical case.

Figure 3.8 shows remarkable agreement between the measured values of v_y and the predictions of the theory given here. It must be remembered that we are using the raw experimental data and there are no adjustable parameters. The values of the material parameters, given in table 3.1, which have been used to normalise the experimental data from the dynamic experiments, have come

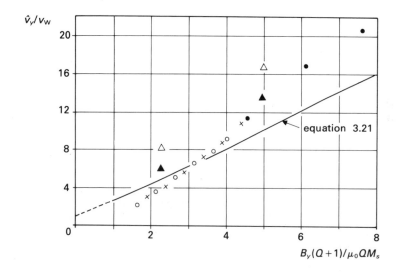

Figure 3.8 The predictions of equation 3.21, using equations 3.19 and 3.20 so that the in-plane field, expressed as $B_y(Q+1)/\mu_0 Q M_s$, is the independent variable, are compared with experimental values of the peak velocity, $\hat{v}_y$, normalised to v_W. Data marked ● are from Rijnierse and de Leeuw (1973), △ and ▲ from de Leeuw and Robertson (1975) and ○ and × from de Leeuw (1977b) and de Leeuw *et al.* (1978). In the last two sets of data △▲ and ○ × distinguish the two possible signs of B_y which were used

from quite independent measurements and have considerable uncertainty. In addition to this, the value of B_y is not easy to determine accurately. Despite these problems, the most recent results agree very well with theory and the earlier results are well within the limits of experimental error. We may conclude that these experiments do not show that the Landau–Lifshitz model is incorrect and that it is a valid and useful theory when it is applied to experiments which satisfy its well defined restrictions. These are that the applied drive fields must not exceed the critical maximum value and that the magnetisation must be kept at a very small angle with respect to the wall plane.

This last point may be underlined if we look back at figure 3.7. When large in-plane fields are applied so that we are working well over to the right-hand side of figure 3.7, the drive field, B_z, cannot be varied over a wide range without the angle ϕ departing well away from zero.

The wall structure can then no longer have any resemblance to the simple Landau–Lifshitz structure and we should not expect the theory given here to apply. This is illustrated in figure 3.9 where the results given by de Leeuw (1973) have been reproduced showing the wall velocities which were observed at drive fields below the drive field required to give the maximum velocity. In this particular case the experimental results appear to lie on a fairly straight line but this line does not pass through the origin. Instead, it extrapolates back to a field of $+3.5 \times 10^{-4}$ T, much too large a field to be explained by any kind of normal

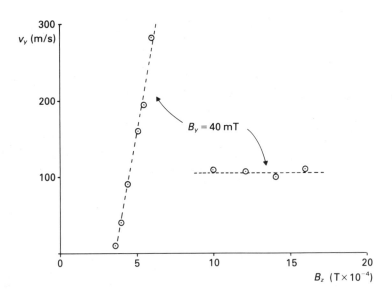

Figure 3.9 The experimental results of de Leeuw (1973) showing the two regimes of wall motion. This is film No. 1 of table 3.1 and the maximum velocity is not shown. Later work by Rijnierse and de Leeuw (1973) gave $\hat{v}_y = 505$ m/s under these conditions

coercivity. A clue to what is happening can be obtained from the slope of this line which is 1.1×10^6 m/s per T. This high apparent mobility, if substituted into equation 3.19 with the values given for $|\gamma|$ and Δ in table 3.1, suggests that $\alpha \approx 10^{-2}$ which is too small by about a factor of 4 in view of the FMR measurements on similar substituted YIG made by Hansen (1974). The slope of the broken line in figure 3.9 is thus unexpectedly high, if we try to apply the Landau–Lifshitz model. What must be happening is that, as the drive field is reduced below the value required for the peak velocity and the angle ϕ becomes negative and large, the exchange and anisotropy fields are no longer in equilibrium with the magnetisation and generate an effective drive field which subtracts from the applied drive field. We can say no more about the behaviour in this low drive region until we know the details of the wall structure which applies. At the higher drives, however, where we observe the maximum velocity, figure 3.8 shows that the problem is understood.

Before we go on to consider what happens at even higher drive fields, where figure 3.9 shows that the wall velocity appears to drop dramatically and become almost drive-independent, we must look at one more point. This concerns the wall screw-sense and is of considerable importance.

3.8 The Importance of the Wall Screw-sense

In equation 3.20, and in the equations which followed, we replaced the symbol B_y by $-|B_y|$ because we had chosen this field to be directed along the negative y-axis so that it balanced the dynamic force which would tend to make M tilt out of the wall plane and take on a positive value of M_y.

This choice of sign came about because we were discussing a wall of the kind first shown in figure 3.1 where the magnetisation lies along the positive x-axis at the exact centre of the wall. This is the wall with the clockwise screw-sense, first discussed at the end of section 2.6. The alternative screw-sense is just as possible: the anticlockwise rotation of M about the y-axis as we move along this axis from one domain to the other.

If we consider the motion of a wall with an anticlockwise screw-sense, we find that M tilts out of the wall plane in the opposite direction to the direction which applies for the clockwise case. To support a high velocity in this kind of wall we need to apply a field along the positive y-axis.

This means that the screw-sense of the moving wall is determined by the relative directions of the velocity vector and the field, B_y. If B_y is antiparallel to v_y the moving wall will have a clockwise screw-sense. If v_y and B_y are parallel, the moving wall will have an anticlockwise screw-sense. This, of course, only applies to walls which are moving near the maximum velocity given by equation 3.21 when the angle ϕ is small and the Landau–Lifshitz wall structure can be assumed to apply. The wall does not have any initial screw-sense, before it begins to move, when these large in-plane fields are applied. The initial structure is of the Néel type (O'Dell, 1978b).

In chapter 2 it was quite clear that the dynamic behaviour of the Landau–Lifshitz wall, moving under the influence of only a B_z field, was quite independent of the sign of its screw-sense. We have now found that the application of the B_y field effectively selects only one of the two possible screw-senses for the moving wall. This has profound implications for bubble domain dynamics because, as we discussed in connection with figure 3.2, there is a built in B_y field for this kind of wall anyway. We can also see that if we have an experiment in which the field B_y is always kept in the same direction and the wall has to reverse its direction of motion for some reason, the screw-sense of the moving wall will also have to reverse. This highly non-linear and almost discontinuous kind of behaviour can produce some very interesting effects which we shall look at later in this chapter.

3.9 The Saturation Velocity

Motion at the peak or maximum velocity, v_y, was discussed in sections 3.6 and 3.7 and figure 3.7 shows clearly how ϕ will reach the unstable value $\hat{\phi}$, unstable because $d\phi/dB_z$ becomes infinite, once B_z exceeds the value

$$\hat{B}_z = \alpha\mu_0 M_s \, (\cos\hat{\phi} \sin\hat{\phi} + b \cos\hat{\phi}) \tag{3.23}$$

which follows when equations 3.19 and 3.20 are substituted into equation 3.14 introducing the Gilbert damping parameter α, equation 2.37, in place of λ.

Figure 3.9 shows some experimental data which illustrate the change in the mode of wall motion that takes place when the drive field is taken above $\hat{B}_z$. As the drive field is increased, there is at first the very rapid increase in the wall velocity, discussed in section 3.7, towards the peak velocity which is, in fact, off the scale of figure 3.9 at 505 m/s. This peak velocity occurs at a drive field around 0.8 mT but at drive fields above this the high velocity mode appears to collapse and the wall moves forward with a fairly drive field independent velocity of 100 m/s, as shown in figure 3.9, between drive fields of 1.0 mT and 1.6 mT. If the values, $\hat{B}_z = 0.8$ mT and $B_y = 40$ mT are substituted into equation 3.23, using equations 3.19 and 3.20, a value of $\alpha = 2 \times 10^{-2}$ is found which is closer to the expected value for this crystal than the $\alpha = 10^{-2}$ deduced from the low field mobility in section 3.7.

This drive-independent wall velocity is often referred to as a saturation velocity and may occur under a variety of experimental conditions. The phenomenon was first discussed at length in papers given by Argyle *et al.* (1971) and Hagedorn (1971) at the 17th MMM Conference in Chicago where the first of a series of papers by Slonczewski (1971, 1972a, 1973) was given which began to clarify the problem.

The reason for a saturation velocity when an in-plane field is applied may be seen by considering figure 3.7. If the in-plane field is high enough to make $b > 1$ it is clear that forward motion of the wall is only possible as ϕ increases from around $-\pi/2$, through zero and then through $\hat{\phi}$. Inertia will maintain this motion as ϕ continues to increase, B_z being in excess of $\hat{B}_z$, but the wall will slow down and must virtually stop as ϕ continues from $+\pi/2$ to $-\pi/2$ again to begin another complete rotation. The angular frequency of this rotation of M in and out of the wall plane, will be $\approx|\gamma|B_z$ which is 2×10^8 rad/s for the B_z of 1.0 mT shown in figure 3.7. The space–time resolution of the equipment used by de Leeuw (1973) would not have been able to resolve this oscillation in the wall velocity and a mean velocity would have been attributed to the wall. We shall see in chapter 4, when the problem of the saturation velocity is discussed from the point of view of the curved magnetic bubble wall, that some experiments have resolved fluctuations of this kind and have already been mentioned briefly in section 1.10.

In other words, this simple model of the saturation velocity has the wall moving forward periodically, the period being $2\pi/|\gamma|B_z$, the velocity being practically zero for half a period and then accelerating up to $\hat{v}_y$ in the next quarter period. During the last quarter period the wall velocity falls back to zero again. It follows that the mean velocity of the wall will not be greater than $\hat{v}_y/4$ and some experimental data is shown in figure 3.10 which confirms this simple picture.

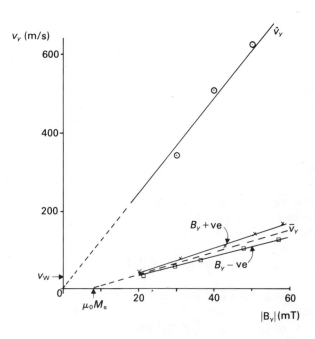

Figure 3.10 Showing how the maximum velocity $\hat{v}_y$, and the saturation velocity, $\overline{v}_y$, depend upon the in-plane field, B_y. These results are from Rijnierse and de Leeuw (1973) for film No. 1 of table 3.1

Figure 3.10 shows some values of $\hat{v}_y$, the peak velocity and $\bar{v}_y$, the saturation velocity, for a range of values of in-plane field B_y. Provided B_y is well in excess of $\mu_0 M_s$, figure 3.10 shows that $\bar{v}_y$ does increase linearly with B_y and that $\bar{v}_y \approx \hat{v}_y/5$.

Quite different behaviour is found as the in-plane field is reduced well below $\mu_0 M_s$ because the saturation velocity begins to vary almost parabolically with B_y (de Leeuw, 1973) and has a well defined value when the in-plane field is reduced to zero. This will be discussed further in chapter 4, because it is of great importance for bubble domain dynamics when the drive field is high. From the point of view of straight domain walls in bubble films, the subject of this chapter, de Leeuw (1978) has made a very detailed study of the experimental data from his own work and from a very large number of other sources. His conclusion was that the saturation velocity under conditions of zero in-plane field was given by $v_W/5$ or, referring to equation 3.18

$$(\bar{v}_y)_{B_y = 0} = |\gamma|\mu_0 M_s \Delta/10 \tag{3.24}$$

This empirical relationship was explained by de Leeuw (1978) in the following way. When there is zero in-plane field and the drive field exceeds the critical value $\alpha\mu_0 M_s/2$, needed to produce the Walker velocity, v_W, the vector M at the centre of the wall region must precess in and out of the wall plane with an angular frequency $\omega_p = |\gamma|B_z$. The magnetostatic energy density of the wall must then be given by $-\frac{1}{2}\mu_0 M_s^2 \sin^2(\omega_p t)$ which represents an average increase of $\mu_0 M_s^2/4$ above the normal magnetostatic energy density, $-\mu_0 M_s^2/2$, of the stationary wall.

This additional energy must be supplied to the wall every half cycle as it moves forward and this means that there must be a power input to the wall

$$P_{IN} = (\mu_0 M_s^2/4)(2h\Delta)(|\gamma|B_z/\pi) \tag{3.25}$$

per unit length, where we have taken the sample thickness to be h and the wall width to be 2Δ.

The power input to the wall can only come from the product of the force upon it, per unit length, and its forward saturation velocity, $\bar{v}_y$. It follows that we also have

$$P_{IN} = 2hM_s B_z \bar{v}_y \tag{3.26}$$

and may equate equations 3.25 and 3.26 to solve for $\bar{v}_y$ as

$$\bar{v}_y = |\gamma|\mu_0 M_s \Delta/4\pi \tag{3.27}$$

which is very close indeed to the empirically established relationship equation 3.24 found by de Leeuw (1978).

A very simple model of the saturation velocity has been given here in which the precession of M about the drive field B_z is assumed to take place with the same phase throughout the entire wall plane. In fact, it is more likely that M will rotate in and out of the wall plane in some kind of progressive manner through the thickness of the film or that there will be different phases in the rotation of M, and thus different instantaneous wall velocities, at different points along the wall. Effects of this kind are of great importance in the theoretical work of Slonczewski (1971, 1972a, 1973) and Hubert (1975) on straight wall motion and its extension by Slonczewski (1974a) and Hagedorn (1974) to the closed bubble domain wall. These problems play a central role in section 4.13.

3.10 Inertial Effects when a Field is Applied Perpendicular to the Wall Plane

We now return to the low drive field regime, where B_z is below the value given by equation 3.23, and consider the problem which was introduced in section 2.14. It was shown there that the domain wall would be expected to have an effective mass because of the way in which the magnetostatic energy of the wall depends upon its velocity.

When we consider the same problem in connection with the kind of experiment we have been discussing in this chapter, we would expect some modification to come into the theory because of the interaction between the wall magnetisation and the large in-plane field which is applied perpendicular to the wall plane. This indeed turns out to be the case but, in what follows, it will become clear that the concept of an effective wall mass is misleading. The very idea of mass belongs to the linear equations of mechanics and, in ferromagnetodynamics, we are dealing with an extremely non-linear problem.

To understand the inertial effects we shall go directly to the equation of motion, as we did in section 2.15. For motion at constant velocity, we previously took equation 3.10

$$\frac{-1}{|\gamma|} \frac{dM}{dt} = (M \times F) - \lambda[F - (F \cdot M)M/M_s^2] \tag{3.28}$$

applied it to the problem, shown in figure 3.5, of a plane wall with the Landau–Lifshitz structure which has fields B_z and B_y applied, and then set the x- and y-components equal to zero because we assumed motion at constant velocity.

When the wall velocity is changing with time the x- and y-components of equation 3.28 are not zero but both yield the relationship

$$\frac{1}{|\gamma|} \frac{d\phi}{dt} = B_z - \mu_0\lambda\{\cos\phi\sin\phi + [|B_y|(Q+1)/\mu_0 M_s Q]\cos\phi\} \tag{3.29}$$

in contrast to equation 3.14 and directly analogous to equation 2.78 which was obtained when we looked at the same problem for the case $B_y = 0$. In equation 3.29 we have again taken the case in which B_y is negative for the reasons which were given after equation 3.20.

Now we know that our equations are only valid when the angle ϕ is small. In the case of constant velocity, this means that B_z and B_y must have values which combine to take us into a region close to the $B_z/\mu_0\lambda$ axis of figure 3.7. If we are outside this region we cannot assume that the wall has the Landau–Lifshitz structure and we have no justification for omitting the exchange and anisotropy fields from our equation of motion 3.28. It follows that if we linearise equation 3.29, by substituting $\phi = \phi_1 + \delta\phi$, where ϕ_1 is a constant angle and $\delta\phi$ is a very small deviation away from ϕ_1, and obtain

$$\frac{d(\delta\phi)}{dt} + \mu_0\lambda|\gamma|\left\{\cos^2\phi_1 - \sin^2\phi_1 - [|B_y|(Q+1)/\mu_0 M_s Q]\sin\phi_1\right\}\delta\phi$$

$$= |\gamma|B_z - \mu_0\lambda|\gamma|\left\{\cos\phi_1 \sin\phi_1 + [|B_y|(Q+1)/\mu_0 M_s Q]\cos\phi_1\right\} \quad (3.30)$$

we are still restricted to small values of ϕ_1 if we are to deduce anything from this result.

Referring to equation 3.14 we recognise that the last term in equation 3.30 is $|\gamma|B_{z_1}$, where B_{z_1} is the drive field which applies for ϕ to take on a steady state value ϕ_1. For the particular case of $\phi_1 = 0$ equation 3.30 is thus simply

$$\frac{d(\delta\phi)}{dt} + \mu_0\lambda|\gamma|(\delta\phi) = |\gamma|(B_z - B_{z_0}) \quad (3.31)$$

showing that, if B_z is reduced or increased slightly away from the value B_{z_0}, at which $\phi = 0$, then $\delta\phi$ falls or rises exponentially with the time constant $(\mu_0\lambda|\gamma|)^{-1}$ which is exactly the same inertial behaviour described by equation 2.82 for the case $B_y = 0$. It may be concluded that the in-plane field has no first order effect on the inertia of the wall, when this is moving at a velocity where $\phi = 0$ and the drive field is changed by a small amount. When ϕ is not zero, however, but has a small value, ϕ_1, equation 3.30 is

$$\frac{d(\delta\phi)}{dt} + \mu_0\lambda|\gamma|\left\{\cos^2\phi_1 - \sin^2\phi_1 - [|B_y|(Q+1)/\right.$$

$$\left.\mu_0 M_s Q]\sin\phi_1\right\}\delta\phi = |\gamma|(B_z - B_{z_1}) \quad (3.32)$$

and the time constant will depend upon ϕ_1. In particular, as ϕ_1 increases from zero towards the maximum possible value, $\hat{\phi}$, given by equation 3.19 and shown in figures 3.6 and 3.7, the second term on the left-hand side of equation 3.32 will get smaller and eventually vanish when $\phi_1 = \hat{\phi}$.

An important conclusion may be drawn from this result. If the wall is moving at the maximum velocity $\hat{v}_y$, which was given by equation 3.21 and shown in figure 3.8 as a function of B_y, and the drive field, B_z, is suddenly reduced to zero, equation 3.32 shows that $\delta\phi$ can only begin to fall at the rather slow rate of $-|\gamma|\hat{B}_z$, where $\hat{B}_z$ is the drive field required for $\hat{v}_y$. Because the wall velocity is determined by the angle ϕ, it follows that the wall must continue to move. Furthermore, it will continue to move at very nearly the maximum velocity for some time because of the steepness of the curves shown in figure 3.7 as they cross the $\phi = 0$ axis. In the experiments we have been discussing in this chapter, $\hat{B}_z$ was typically 5×10^{-4} T so that $|\gamma|\hat{B}_z$ is of the order of 10^8 rad/s and we would expect the peak velocity to persist for several nanoseconds if an experiment could be made in which the drive field were suddenly reduced to zero.

Experiments showing inertial effects of this kind, in cases where a large in-plane field is applied perpendicular to the plane of the moving wall, were described by de Leeuw and Robertson (1975) and, in more detail, by de Leeuw *et al.* (1978). In these experiments, a pulsed drive field, just below the value $\hat{B}_z$, was applied to move the wall which was stabilised by means of an externally applied gradient field, or quadrupole field, as we described in section 3.4. The result was that the wall moved to a new equilibrium position where B_z was again zero. The drive field is thus being reduced to zero in these experiments, not suddenly but as the wall moves. The expected overshoot was clearly observed and the wall response was highly non-linear, as the very non-linear equations of motion would lead us to expect. The wall position underwent triangular oscillations, the decay of these oscillations being almost linear with time, not exponential as a linear oscillator would be. The frequency of the oscillations was certainly consistent with the model described above in that it was of the order of $|\gamma|\hat{B}_z$, $\hat{B}_z$ increasing with B_y.

Equation 3.32 is already in a form which could be solved numerically, step by step, to tell us how ϕ relaxes from $\hat{\phi}$, through zero, to then begin the rapid reversal of the wall screw-sense which was discussed in section 3.8 and which must take place as the wall stops and starts to move in the opposite direction. During the reversal of the screw-sense, however, ϕ must take on a large value and our equations do not cover this case. The motion is highly non-linear and any attempt to force the experimental results to fit a linear oscillator model, bringing in the concept of an equivalent wall mass, is just misleading. This criticism does not apply quite so definitely to a rather similar kind of wall motion which we shall now consider briefly.

3.11 Wall Motion when an In-plane Field is Applied Parallel to the Wall

A number of experiments were referred to in section 3.11 in which the dynamics of straight domain walls were studied using stripe domain patterns in bubble domain films. In these experiments a straight wall is stabilised by using the quadrupole field generated by its two neighbouring walls, instead of using an externally applied gradient field or quadrupole field of the kind shown in figure 3.4. Glancing back at figure 3.2 shows that such a field can be produced in this way if we imagine two walls, both of the opposite polarity to the one shown in figure 3.2, lying on either side of the wall originally shown there.

Such a stripe domain pattern is still susceptible to very long wavelength instabilities and may degenerate into the well known serpentine pattern, typical of the demagnetised state in a magnetic bubble film. This instability can be suppressed if a magnetic field is applied in the plane of the sample and in the direction along which the straight walls are required. It was natural, in view of this experimental requirement, that the dynamics of the straight stripe domain walls were investigated when this in-plane field was applied. This was done by Argyle and Malozemoff (1972), Argyle and Halperin (1973), Moody *et al.* (1973) and by Shaw *et al.* (1974).

The results of these experiments appear to be very similar to those we have discussed in this chapter where the in-plane field is applied perpendicular to the wall plane. Very high wall velocities were observed and wall oscillations, again with the distinctive triangular waveform, were also seen.

Before looking at the equations of motion for this new situation we must consider the static wall structure when the in-plane field is applied. Referring to figure 3.1, let us suppose that the in-plane field, B_x, is positive. It will then support a wall of the clockwise screw-sense, like the one shown in figure 3.1, but tend to make a wall with an anticlockwise screw-sense unstable. The magnetisation within the domains themselves, shown in figure 3.1, will now take up an angle θ_0, given by

$$\sin \theta_0 = B_x/\mu_0 M_s Q \tag{3.33}$$

with respect to the z-axis in (x, z). This result follows from exactly the same argument as the one which led to equation 3.8.

The static structure of the wall itself is considered by O'Dell (1978a). We find that the application of the field B_x makes very little difference to the original Landau–Lifshitz structure provided $B_x/\mu_0 M_s Q \ll 1$. As we would expect, the wall is widened very slightly and we can include the effect of this by bringing θ_0 into the z-component of the Landau–Lifshitz equation

$$\frac{1}{|\gamma|} \frac{dM}{dt} = (M \times F) - \lambda [F - (F \cdot M)M/M_s^2] \tag{3.34}$$

by writing

$$\frac{dM_z}{dt} = (M_s v_y/\Delta)\cos \theta_0 \tag{3.35}$$

in analogy to equation 3.15, where we are again concentrating on the centre of the wall so that

$$M_x = M_s \cos \phi$$

$$M_y = M_s \sin \phi \qquad (3.36)$$

$$M_z = 0$$

and

$$F_x = \mu_0 M_s \cos \phi + B_x$$

$$F_y = 0 \qquad (3.37)$$

$$F_z = B_z$$

We consider motion at constant velocity first, in order to find out what the effect of the field B_x is. The x- and y-components of equation 3.34 vanish when v_y is constant and both give the relationship

$$B_z = \mu_0 \lambda [\cos \phi \sin \phi + (B_x/\mu_0 M_s)\sin \phi] \qquad (3.38)$$

between the applied fields and the angle ϕ which M makes with the moving (x, z) of the wall.

Using equation 3.35, the z-component of equation 3.34 is

$$(v_y/|\gamma|\Delta)[1 - (B_x/\mu_0 M_s Q)^2]^{1/2} =$$

$$\mu_0 M_s \sin \phi \cos \phi + B_x \sin \phi - (\lambda/M_s)B_z \qquad (3.39)$$

where we have used equation 3.33.

Equation 3.38 may now be substituted into equation 3.39 to give the relationship between the wall velocity and the applied fields as

$$v_y = \frac{|\gamma|\Delta M_s(1 + \lambda^2/M_s^2)}{\lambda[1 - (B_x/\mu_0 M_s Q)^2]^{1/2}} B_z \qquad (3.40)$$

which is very similar to our previous result, equation 3.17, for the case where the in-plane field is applied perpendicular to the wall plane.

There is, however, a very considerable difference between these two cases. When the in-plane field was applied perpendicular to the wall plane we found that the angle ϕ could remain very small at high velocity and we could expect the model to be a fairly accurate one. Equation 3.38, on the other hand, shows that B_z, and thus the wall velocity, has a maximum value when

$$\phi = \hat{\phi} = \arccos\left\{ [-c + (c^2 + 8)^{1/2}]/4 \right\} \qquad (3.41)$$

where

$$c = B_x/\mu_0 M_s Q \qquad (3.42)$$

Equation 3.41 shows that the effect of the field B_x is to allow ϕ to take on a maximum value greater than $\pi/4$, the value which applies for the Walker velocity when $B_x = 0$. This is a very artificial result, however, because it contradicts our requirement that ϕ must remain small if we are going to be able to continue ignoring the exchange and anisotropy fields.

Our conclusion is that the effect of the field B_x is similar to the previous case for the field B_y in that a higher wall velocity is obtained, for the same value of the angle ϕ, than when these fields are absent. Comparing equations 3.14 and 3.38, however, shows that the B_y field is far more effective when ϕ is small because it is associated with $\cos\phi$, whereas B_x is multiplied by $\sin\phi$ in equation 3.38.

Another difference between the two cases is the connection between the wall screw-sense and the wall velocity when the in-plane field is applied. A simple argument from the point of view of symmetry would lead us to expect the connection discussed in section 3.8 when the in-plane field is perpendicular to the wall plane. Similarly, we would not expect to find such a connection in the case where the in-plane field is applied parallel to the wall and equation 3.38 confirms this. The angle ϕ is zero when B_z is zero and changes sign with B_z. There is, of course, a connection between the screw-sense of the wall and the sign of B_x but this is a static relationship and has been discussed above.

Finally, let us look at the problem of wall acceleration when an in-plane field is applied parallel to the wall. The x- and y-components of equation 3.34 then no longer vanish but give

$$\frac{d\phi}{dt} = |\gamma|B_z - \mu_0\lambda|\gamma|[\cos\phi\,\sin\phi + (B_x/\mu_0 M_s)\sin\phi] \qquad (3.43)$$

as the differential equation relating ϕ and the applied fields. It is only possible to discuss this equation for small values of ϕ, again for the reasons mentioned above, and it is then linear

$$\frac{d\phi}{dt} + \mu_0\lambda|\gamma|[1 + (B_x/\mu_0 M_s)\phi] = |\gamma|B_z \qquad (3.44)$$

An effective wall mass may now be found by using equation 3.39, again for small ϕ, to obtain the first order differential equation in velocity as

$$\frac{2(1-c^2)^{1/2}}{\mu_0\gamma^2\Delta(1+B_x/\mu_0M_s)}\frac{dv_y}{dt}+\frac{2\lambda(1-c^2)^{1/2}}{|\gamma|\Delta}v_y = 2M_sB_z \qquad (3.45)$$

where we have used equation 3.42 for conciseness.

Equation 3.45 should be compared with equation 2.83 which arose when we discussed the effective mass of the simple Landau–Lifshitz wall. It is then clear that the effective mass of the wall is reduced, when an in-plane field is applied parallel to the wall plane, in that equation 3.45 gives

$$m = (2/\mu_0\gamma^2\Delta)[(1-c^2)^{1/2}/(1+B_x/\mu_0M_s)] \qquad (3.46)$$

whereas the Döring mass, derived in chapter 2 as equation 2.77, was

$$m_D = 2/\mu_0\gamma^2\Delta \qquad (3.47)$$

It follows that, if we could measure the effective mass as a function of the in-plane field B_x, we should find a linear relationship between m^{-1} and B_x

$$m^{-1} = m_D^{-1}(1+B_x/\mu_0M_s) \qquad (3.48)$$

because in most practical cases $c^2 \ll 1$, in equation 3.46, and may be neglected. Such a linear relationship between m^{-1} and B_x was observed by Shaw *et al.* (1974) but m^{-1} increased far more rapidly than equation 3.48 would predict. The reason for this may be seen from the form of the wall transient response which was observed in these experiments. This is clearly non-linear, the wall position oscillating with a triangular waveform, not a sinusoid, and the amplitude of these oscillations decaying linearly, not exponentially. The drive field which was used in these experiments was 3.4×10^{-4} T and as $\alpha = \lambda/M_s$ for these materials is certainly below 10^{-2}, equation 3.38 shows that a drive field of this magnitude is far too large to allow the approximation which led to equation 3.44. An effective wall mass has no meaning when a linear equation of motion cannot be assumed.

3.12 Conclusions

In this chapter a dynamic experiment has been considered which is particularly well defined: it involves a moving wall which is straight, can be observed magneto-optically and also the wall structure may be inferred with some confidence. The range of applied fields and wall velocities which become available to the experimentalist, under the conditions described in this chapter, has been greatly extended in comparison to the work described in chapter 2. This comes about not only because of the applied in-plane field but also because of the special nature of the materials being used. These are the bubble domain garnet layers which have a very low value of $\mu_0 M_s$, compared to the materials discussed in the previous chapter, so that even though the observed wall velocities may extend up to ten times the Walker velocity, $v_W = |\gamma|\mu_0 M_s \Delta/2$ (equation 3.4), these will still be well below the critical velocity, $v_c = |\gamma| \sqrt{(\mu_0 A)}$ (equation 2.109).

The problem of the applied drive field causing the moving wall to bend was also discussed in chapter 2. This problem does not arise in the experiments described in this chapter because the bubble garnet layers have a characteristic length, l, comparable with their thickness, h, so that the field $\hat{B} = \mu_0 M_s l/h$ (equation 2.119), is large. In any case wall bending of the kind discussed in section 2.18 would occur along the length of the moving wall in a bubble garnet layer, not through the thickness of the layer, and even in the very high drive field experiments (de Leeuw, 1977b) wall bending was not observed in the streak camera photographs which were used to determine the wall velocity.

The majority of bubble domain garnet layers show some kind of in-plane anisotropy and an interesting example of how this may affect the dynamic properties has been given by Malozemoff and Papworth (1975) using a sample in which an additional in-plane anisotropy had been introduced by growing the expitaxial garnet layer upon a misaligned substrate. The problem is very similar to the one considered in this chapter except that the anisotropy field always has the same sign as the magnetisation so that the screw-sense selection effects, discussed in section 3.8, do not arise. In some materials, prepared in the form of thin layers, the effects of the in-plane anisotropy may be very large and the best examples of such materials are found among the orthoferrites. Some experiments on bulk crystals of these interesting materials were described in section 1.4. It is clearly questionable to treat these canted sublattice, or weakly ferromagnetic, materials as simple ferromagnets with isotropic exchange. This has been discussed by Joenk (1964) and by Gyorgy and Hagedorn (1968) who concluded that a model in which the weak ferromagnetic moment would rotate about an axis perpendicular to the plane of the moving wall would be valid. The remarks made in sections 2.7 and 2.18 concerning the wall structure proposed by Bulaevskii and Ginzburg (1964, 1970), which may well apply at normal temperature for many of these orthoferrite materials, which can have very high values of Q, should be born in mind, however.

The idea of saturation velocity for the domain wall has been introduced in this chapter. This can lead to some very interesting ideas when the moving wall is no longer straight but is the closed wall of a moving bubble domain. Saturation will be considered in section 4.13, and a good bridge between the work described in this chapter and that in the next chapter has been made by de Leeuw (1977a) in his review of wall dynamics in bubble domain garnet layers.

The interesting inertial effects which were discussed in the closing sections, 3.10 and 3.11, of this chapter have been presented in a slightly different way elsewhere (O'Dell, 1979) to emphasise the large mass which might be associated with equation 3.32 when the drive field is very close to the value which should give the peak wall velocity. Under such conditions equation 3.32 tends to $d(\delta\phi)/dt = 0$ which implies a very large inertia. Effects of this kind would depend upon the dynamic wall structure remaining stable as the wall moves forward and, again it is in bubble domain dynamics that we may see experimental evidence for dramatic changes in wall structure being brought about by wall motion.

4 Magnetic Bubble Domain Dynamics

4.1 Introduction

The magnetic bubble domain was introduced in figure 1.10 as an isolated cylindrical domain passing through a thin layer of magnetic material which has an easy direction of magnetisation perpendicular to its plane. The bubble domain is supported by an externally applied bias field which must lie within a critical range if the bubble is neither to collapse nor to run out into a strip domain.

The history of the experimental work on magnetic bubble domain dynamics was given briefly in the final sections of chapter 1. This chapter begins with a similar account of the theoretical development of bubble dynamics in order to identify the main sources in the literature. In the sections which follow, however, we shall not develop the theory along the same track which it took historically because, as always, problems are clearer with hindsight.

4.2 Historical Development

The first description of the potential data processing applications of magnetic bubbles was given by Bobeck (1967) but dynamic problems were not mentioned until Perneski (1969) began to discuss the motion of a magnetic bubble in an externally applied magnetic field gradient and used the relationship

$$v_n = \mu_w B_z \qquad (4.1)$$

referring to work which was published later by Thiele (1971).

Equation 4.1 simply defines a linear relationship between the normal component of the wall velocity, v_n, and the drive field, B_z, by introducing the idea of a constant wall mobility, μ_w. The same concept was discussed in chapter 1, section 1.3, and belonged to the earlier times of ferrite core stores and thin permalloy films. Nevertheless, much interesting work was published in bubble dynamics using this simple wall mobility model.

Callen and Josephs (1971) used the constant mobility model to give an interpretation of the bubble collapse experiment, first described by Bobeck et al. (1970). Callen et al. (1972) published a very interesting paper which described a phenomenological model for bubble dynamics involving a velocity dependent wall mobility and wall mass. Cape (1972), in a paper which was specifically devoted to bubble dynamics, also developed a phenomenological model but this introduced some ideas from a paper by Henry (1971) which, along with some very brief notes by Slonczewski (1971) (Argyle et al., 1971), really mark the beginnings of an atomic scale model for magnetic bubble domain dynamics as opposed to the simple wall mobility model or the phenomenological models which had been used up to then. The phenomenological approach was brought to a very high level of sophistication, however, in a very interesting paper by Cape *et al.* (1974).

It became clear that an atomic scale model was really essential for magnetic bubble domain work when the hard magnetic bubble was discovered. The experimental discovery of the hard bubble was described in section 1.9, and the first attempts to explain hard bubble dynamics (Vella-Coleiro *et al.*, 1972b), Malozemoff and Slonczewski, 1972, Slonczewski, 1972b and Thiele, 1973b have been discussed by Thiele *et al.* (1973). In parallel with this, a foundation paper by Slonczewski (1972a) led to a series of most interesting papers (Slonczewski, 1973, 1974a, 1974b) dealing with more complicated atomic scale wall structures and their dynamic properties. New developments in the theory of magnetic domain wall dynamics were also being published by Thiele (1973a, 1974, 1975, 1976), Hagedorn (1974), Hubert (1975), Nedlin and Shapiro (1975, 1977) and Schlömann (1975, 1976). All this work was clearly instigated by the problems coming from bubble domains.

Notable steps forward were also made towards solving some of the computational problems of bubble domain dynamics. Schryer and Walker (1974) went very thoroughly into the problem of solving the equations of domain wall motion digitally, but their work was only concerned with straight walls and, as we shall see, an essential feature of bubble dynamics is that the closed domain wall may conserve any structure which the wall has along its length. Hubert (1975) considered the computational problems of bubble domain walls in great detail but was only able to include the possibility of wall structure through the thickness of the wall and the thickness of the bubble film. An attempt to consider real bubble dynamics was made by Hayashi and Abe (1976) in a very interesting paper which was a development of earlier work using the constant mobility model, equation 4.1, Hayashi and Abe (1975).

The theory of magnetic bubble domain dynamics is concerned with the response of magnetic bubbles to time varying bias fields, causing changes in bubble diameter, and to bias fields which vary spatially, causing bubble translation. The theory is complicated, when compared to the methods of chapters 2 and 3 for straight walls, because the bubble domain has a closed wall: the closed wall of a cylinder (θ, z). Even when it is possible to ignore any variation in the wall structure through the thickness of the bubble film, z, it is not possible to neglect the θ-dependence of the bubble wall structure because this may be effectively trapped within the closed wall.

The possibility of a θ-dependent bubble wall structure is suggested first of all by the static behaviour of the so-called hard bubble domain and we shall look at this interesting static problem by way of introduction to the work on bubble dynamics. Before doing this, however, a brief review must be given of the static properties of what are known as normal bubbles in order to define the usual parameters and give some fundamental relationships for magnetic bubble domains.

4.3 The Statics of Normal Magnetic Bubbles

The static relationship between the externally applied bias field and the diameter of the normal bubble domain, in terms of the material parameters and the thickness of the layer which supports the bubble, has been given by a number of authors since the problem was first discussed by Bobeck (1967) and treated in depth by Thiele (1969, 1970, 1971). The problem is also treated in three textbooks on magnetic bubbles: O'Dell (1974), Bobeck and Della Torre (1975) and Chang (1975), the last of these being more concerned with applications and having an excellent collection of forty-six foundation papers reprinted as appendices. A more recent book by Chang (1978) is specifically concerned with applications.

The so-called 'normal bubble' or 'soft bubble' is assumed to have a simple wall structure of the kind shown in figure 4.1. This structure does not depend upon θ and is assumed to vary according to the Landau–Lifshitz solution of chapter 2 (equations 2.30, 2.31), as we cross the wall region near the plane $z = 0$. As we move through the thickness of the film, from $z = -a$ to $z = +a$, the wall is assumed to have the z-dependence shown in figure 4.1, first discussed here in chapter 3, figure 3.3.

The magnetic bubble will be in static equilibrium when the net magnetic field at the wall is zero. There are three fields to take into account: the bias

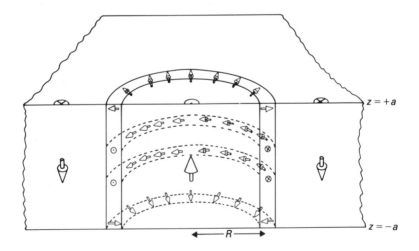

Figure 4.1 The wall structure of the normal or soft bubble approximates towards the Landau–Lifshitz structure near the plane $z = 0$ but has a Néel structure near the top and bottom surfaces of the thin bubble film, $z = \pm a$

field, the magnetostatic field of the bubble itself and an additional exchange field which comes in because the bubble domain wall is curved. We are only concerned with the z-components of these fields because only a z-component can produce a force upon the wall.

Let us consider the additional exchange field first. The exchange field was introduced in chapter 2, equation 2.15 as

$$B_{ex} = (2A/M_s^2) \nabla^2 M \tag{4.2}$$

and we are only concerned with its z-component which will be given by

$$(B_{ex})_z = (2A/M_s^2)\left(\frac{1}{r}\frac{\partial}{\partial r}\left(r\frac{\partial M_z}{\partial r}\right)\right) \tag{4.3}$$

in cylindrical polar coordinates because M_z is a function of r only for the structure shown in figure 4.1.

Equation 4.3 is the sum of two terms

$$(B_{ex})_z = (2A/M_s^2)\left(\frac{1}{r}\frac{dM_z}{dr} + \frac{d^2 M_z}{dr^2}\right) \tag{4.4}$$

to be evaluated at the bubble wall, $r = R$. The second term in equation 4.4 is the familiar exchange term, corresponding to $(2A/M_s^2)\,d^2 M_z/dy^2$ in the planar wall problem of chapter 2. The first term in equation 4.4 is the new factor which depends upon the wall curvature and is usually called the effective wall field. If we assume that the wall structure is approximately given by the Landau–Lifshitz solution, equation 2.30 of chapter 2, then $dM_z/dr = -M_s/\Delta$ at $r = R$ and we may write the effective wall field as

$$B_w = -2A/M_s R\Delta \tag{4.5}$$

a field which is opposed to the direction which M_z has inside the bubble and thus tends to make the bubble domain contract, as does the bias field B_0. It is the sum of B_0 and B_w which must be in equilibrium with the magnetostatic field of the bubble which we shall denote by $\bar{B}$ because we need a mean value, over the interval $-a < z < +a$, of the magnetostatic field evaluated at $r = R$. It may be shown (O'Dell, 1974) that $\bar{B}$ is quite accurately represented by the expression

$$\bar{B} = \mu_0 M_s[1 + 0.726(R/a)]^{-1} \tag{4.6}$$

which is a modified version of an approximation given by Callen and Josephs (1971).

Normalising all fields to $\mu_0 M_s$, it is clear that the equilibrium condition for the normal bubble domain will be given when

$$B_0/\mu_0 M_s + B_w/\mu_0 M_s + [1 + 0.726(R/a)]^{-1} = 0 \qquad (4.7)$$

Using the parameters, $Q = 2K_u/\mu_0 M_s^2$, $\Delta = (A/K_u)^{1/2}$ and introducing the material parameter, l, where

$$l = 2Q\Delta = 4(AK_u)^{1/2}/\mu_0 M_s^2 \qquad (4.8)$$

is called the characteristic length, a very common parameter in bubble domain work (O'Dell, 1974), we can substitute equation 4.5 into equation 4.7 and write the normal bubble equilibrium condition as

$$|B_0|/\mu_0 M_s = [1 + 0.726(D/h)]^{-1} - l/D \qquad (4.9)$$

using the diameter of the bubble, $D = 2R$, as a more convenient variable and writing $h = 2a$.

Equation 4.9 has two roots for D in any given bias field below the collapse field but only the larger value of D is stable. The widest range of bubble diameters is obtained when $l/h \approx 0.2$, and this is the usual choice when the material is intended for device work (Thiele, 1971).

Equation 4.9 shows that the stable bubble diameter decreases as the bias field is increased and the bubble should collapse when D falls to its collapse value

$$D_{col} = 1.174(lh)^{1/2}/[1 - (0.726l/h)^{1/2}] \qquad (4.10)$$

which is easily found when we differentiate equation 4.9 with respect to D and set $dB_0/dD = 0$.

The theory given above is compared with some experimental data in figure 4.2 where equation 4.9 has been plotted on the left, labelled 'normal bubble characteristic'. The bubble film being used in this case had $l = 1.08 \mu m$, $h = 6.05$ μm and $\mu_0 M_s = 11.7 mT$ and the values of D, from equation 4.9 have been normalised to D_{col}, from equation 4.10, because this makes it far easier to present the experimental data accurately.

Good agreement between measured values of bubble diameter, taken over a range of bias field, and the normal bubble characteristic, shown in figure 4.2, would be obtained with any good quality garnet film of the kind which would be intended for use in a bubble domain device. In the case of the bubble film being discussed here, however, one vital step in the process of its preparation was omitted. This step is called ion-implantation and will be discussed in detail in section 4.8. Let us accept for the moment that, because the ion-implantation step has been omitted, this film will show the interesting hard bubble phenomena

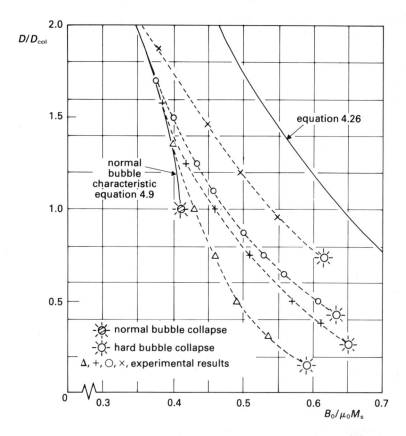

Figure 4.2 Hard bubble diameter against field measurements taken on a $(CaSm_{0.1}Y_{1.9})(GeFe_4)O_{12}$ LPE film 6.05 μm thick having $\mu_0 M_s = 11.7$ mT, $l = 1.08$ μm and $Q = 10.3$

which are the subject of the next section. Our first problem is to understand the static properties of these so-called hard bubbles.

4.4 Hard Bubble Statics

The contrast between the static behaviour of normal bubbles and hard bubbles in a typical device film which has not been ion-implanted is shown in figure 4.2. Here, some experimental results are shown of bubble size against bias field. The bubble domain diameter is normalised to the normal bubble collapse diameter, equation 4.10, because this is also measured experimentally and is subject to the same absolute error as the other diameter measurements. Four separate sets of data are shown and these come from four different bubble domains, nucleated in the film by the usual stripe chopping technique. In all four cases the bubble diameter–bias field characteristic lies well to the right of the normal bubble characteristic and terminates in what we have labelled 'hard bubble collapse'. Three distinct points may be seen at once concerning these hard magnetic bubbles: in all four cases, bubble collapse occurs at higher bias field than normal collapse, at a smaller bubble diameter than the normal bubble collapse diameter and, finally, the slope of the bubble diameter–field characteristic is quite small just before collapse in comparison to the very high value of dD/dB_0 which the normal bubble characteristic has just before the curve reaches normal bubble collapse. This last point means that hard bubble collapse must take place by means of some structural change as opposed to normal collapse which is just a simple instability.

The three points discussed above were taken by Tabor *et al.* (1972a, 1972b) and by Malozemoff (1972) as evidence that the hard bubble must have a wall structure of the kind shown in figure 4.3. The magnetisation shown in figure 4.3 not only rotates in the plane of the wall, (θ, z), as we go from the inside of the bubble to the outside but also rotates in the plane of the film, (r, θ), as we go around the wall circumference.

As shown in figures 4.3a and b, there are two distinct forms of this wall structure, both having the same number of alternations of M_θ around the circumference but different senses of rotation for M_θ in (r, θ). These topological properties of the hard bubble wall structure were discussed by Slonczewski *et al.* (1972) and by Voegeli and Calhoun (1973) who proposed the now universally adopted S-number description of the bubble wall. The S-number, or state number, is a measure of how many times M rotates in (r, θ) as we travel once around the

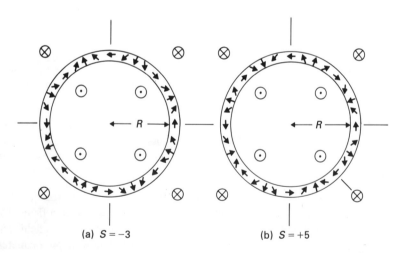

(a) $S = -3$ (b) $S = +5$

Figure 4.3 Showing the kind of wall structures which have been proposed to explain hard bubble behaviour. For clarity, the wall width has been exaggerated and only four alternations in M_θ are shown around the circumference. A really hard bubble would have more than ten times this number

bubble wall. The sign of S is positive when M rotates in (r, θ) in the same sense as the sense chosen for travel around the bubble wall so that the bubble wall shown in figure 4.3a must be labelled $S = -3$ while that shown in figure 4.3b must be labelled $S = +5$.

We shall see at the end of this section that really hard bubbles have $|S| \approx 100$ so that the bubble wall structures shown in figure 4.3 are only illustrations of the idea being introduced, the diagrams appearing fairly simple because the wall width has been exaggerated relative to the bubble diameter. Bubbles having the very low S-numbers shown in figure 4.3 would, in fact, have a rather complicated asymmetry which will be considered in section 4.9. In this section we are only concerned with very hard bubbles and this is an important point because when the hard bubble wall structure is what we shall call tightly wound, it is possible to neglect the magnetostatic interactions around the wall and also the effect of the radial magnetostatic field at the top and bottom surfaces of the bubble film which, in the normal bubble (Della Torre *et al.*, 1975, Hubert, 1975) causes M to point along r close to the film surface. This follows because exchange completely dominates the problem when the wall structure is tightly wound.

Before proceeding, the term 'Bloch line' which has come into the hard bubble literature should be mentioned. This term was first introduced (DeBlois and Graham, 1958) to describe a region where the screw-sense of a domain wall changes sign in a thin film of some isotropic material, like permalloy, where the magnetisation would normally lie in the plane of the film. Such transitions in the wall screw-sense were studied experimentally by Methfessel *et al.* (1960) using colloid techniques on permalloy films. The problem is very different in a magnetic bubble domain film, however, where the magnetisation within the domains is normal to the plane of the film. In a bubble film, div M must become very large around the region where the wall screw-sense changes sign. This was discussed by Feldtkeller (1965), in a theoretical paper long before the advent of bubble domains, and he suggested that the term 'Néel line' should be used to distinguish this case from the Bloch line which had become familiar to workers with permalloy films. This advice was not taken, however, and the expressions 'vertical Bloch line' and 'horizontal Bloch line' were introduced (Malozemoff and Slonczewski, 1972) to describe a change in the sign of the wall screw-sense with an increase in θ or in z, respectively. For example, the bubbles illustrated in figure 4.3 would both be said to have eight vertical Bloch lines. These expressions will be avoided in what follows and the term Néel line used only when this can aid conciseness.

To return to our discussion of the static properties of hard magnetic bubbles, the wall structures shown in figure 4.3 clearly explain some of the observed behaviour qualitatively. As the bias field is increased the wall structures shown in figure 4.3 will have to become more tightly wound as the bubble diameter decreases. This must increase the exchange energy and if this increases rapidly enough there will be no reason for the hard bubble to collapse. The hard bubble should, in fact, simply get smaller and smaller as the bias field is increased and this is why we find a finite slope for the hard bubble characteristics, shown in figure 4.3, persisting right up to hard bubble collapse. Collapse itself can only come about by some sudden breakdown of the exchange coupling which allows the whole structure to unwind. As Hagedorn (1974) has mentioned, the details of this unwinding have never been explained. The finite temperature of the spin lattice or, perhaps, point defects in this lattice must play an important role (Haisma *et al.* 1975).

Quantitatively, we may express the problem very simply if we use a local cartesian coordinate system (y, x, z) on the wall regions shown in figure 4.3, in place of (r, θ, z), and neglect the wall curvature. Distances along the circumference of the bubble are then given by the local coordinate x. A representation of M within the wall which gives the required constant amplitude to M is then

$$M_x = \pm M_s \operatorname{sech}(y/\delta y) \sin(\pi x/\delta x) \qquad (4.11)$$

$$M_y = \pm M_s \operatorname{sech}(y/\delta y) \cos(\pi x/\delta x) \qquad (4.12)$$

and

$$M_z = \pm M_s \tanh(y/\delta y) \qquad (4.13)$$

The choice of signs in equations 4.11 and 4.12 now comes from the choice in the orientation of the patterns shown in figure 4.3 and the choice in the sign of the S-number. The two signs in equation 4.13 simply reflect the choice of sign of M_z within the bubble domain. Equations 4.11 to 4.13 are extensions of the Landau–Lifshitz wall discussed in chapter 2, equations 2.30 and 2.31.

In equations 4.11, 4.12 and 4.13 δy corresponds to the wall width parameter, $\Delta = (A/K_u)^{1/2}$, of the Landau–Lifshitz wall but we shall find that δy is not a constant, although we shall have to assume that it is independent of x. The parameter δx is the distance which we must move around the circumference of the bubble in order to observe one reversal in the direction of M_θ or M_r. It follows that

$$\delta x = \pi D/2n \qquad (4.14)$$

where n is the number of complete cycles which M makes as we travel around the complete wall circumference. In both figure 4.3a and figure 4.3b, $n = 4$.

We must now find the total energy density of this hard bubble wall structure and this will be given by the sum of the exchange energy density

$$E_{\text{ex}} = -(A/M_s^2)M \cdot \nabla^2 M \qquad (4.15)$$

which was derived in chapter 2 as equation 2.13, and the anisotropy energy density, $K_u \sin^2 \phi$, where ϕ is the angle which M makes with the easy axis, z.

It is clear from equations 4.11 and 4.12 that $\sin \phi \equiv \text{sech} \,(y/\delta y)$ so that we have a total energy density given by

$$E = -(A/M_s^2)M \cdot \nabla^2 M + K_u \,\text{sech}^2 (y/\delta y) \qquad (4.16)$$

The total energy per unit area of the hard bubble wall may now be found by substituting equations 4.11, 4.12 and 4.13 into equation 4.16, using equation 4.14 to express δx in terms of D and n, expressing K_u in terms of Δ and integrating with respect to y using the rather artificial, but nevertheless accurate, limits $-\infty$ to $+\infty$. The result is

$$\sigma_{whb} = 2A\,[(2n/D)^2 + (1/\delta y)^2 + (1/\Delta)^2]\,\delta y \qquad (4.17)$$

which is a minimum, for given values of n, D and Δ when

$$\delta y^2 = \Delta^2/[1 + (2n\Delta/D)^2] \qquad (4.18)$$

which follows from simply differentiating equation 4.17 with respect to δy.

Equation 4.18 shows that the width of the hard bubble wall tends to shrink as the bias field is increased and D gets smaller. If equation 4.18 is now put back into equation 4.17 we shall have σ_{whb} as a function of D, the material parameters and n only, that is

$$\sigma_{whb} = (4A/\Delta)[1 + (2n\Delta/D)^2]^{1/2} \qquad (4.19)$$

and the total energy of the hard bubble wall, length πD and height $h = 2a$, is thus

$$E_{whb} = (4\pi Ah/\Delta)(D^2 + 4n^2\Delta^2)^{1/2} \qquad (4.20)$$

Now the simplest way to obtain the relationship between the hard bubble diameter and the applied bias field is to differentiate E_{whb} with respect to $D/2$, and thus obtain

$$f_{whb} = -8\pi ADh/(D^2 + 4n^2\Delta^2)^{1/2}\Delta \qquad (4.21)$$

as the contracting force due to the hard bubble wall energy. This may then be expressed as an effective wall field, B_{whb}, to replace equation 4.5 in our previous analysis of the normal bubble, by arguing that f_{whb} must be given by

$$f_{whb} = (2hM_s)(B_{whb})(\pi D) \qquad (4.22)$$

(O'Dell, 1974). Equation 4.21 is then combined with equation 4.22 and 4.8 to give

$$B_{\text{whb}}/\mu_0 M_s = -l/(D^2 + 4n^2 \Delta^2)^{1/2} \qquad (4.23)$$

Equation 4.23 may be compared with an expression for the normal bubble effective wall field which may be obtained by substituting equation 4.8 into equation 4.5 to give

$$B_w/\mu_0 M_s = -l/D \qquad (4.24)$$

which was, of course, the term that finally appeared at the end of equation 4.9. The hard bubble diameter–field characteristic is now obvious because all that need be done is to replace the last term in equation 4.9 with the right-hand side of equation 4.23 to obtain

$$|B_0|/\mu_0 M_s = [1 + 0.726(D/h)]^{-1} - l/(D^2 + 4n^2 \Delta^2)^{1/2} \qquad (4.25)$$

Equation 4.25 makes it clear why the hard bubble characteristics shown in figure 4.2 tend to follow the shape of the characteristic for zero effective wall field, that is the curve

$$|B_0|/\mu_0 M_s = [1 + 0.726(D/h)]^{-1} \qquad (4.26)$$

as we approach hard bubble collapse. The reason must be that D^2 has fallen well below $(2n\Delta)^2$ and equation 4.25 reduces to

$$|B_0|/\mu_0 M_s = [1 + 0.726(D/h)]^{-1} - l/2n\Delta \qquad (4.27)$$

which is simply equation 4.26, the heavy curve on the far right-hand side of figure 4.2, shifted over to the left by a distance $l/2n\Delta$. This is easier to think about quantitatively if we use equation 4.8 to write $l/2n\Delta = Q/n$.

The hardest of the four bubble domains represented in figure 4.2 has a characteristic which lies about 0.1, on the normalised bias field scale, to the left of the zero wall field characteristic. It follows that $Q/n \approx 0.1$ for this particular hard bubble and, as $Q = 10.7$ for this film we must have $n \approx 100$. The other hard bubbles represented in figure 4.2 have smaller values of n; the two whose characteristics lie close together both have $n \approx 50$. The remaining hard bubble, which has the smallest collapse diameter, has $n \approx 30$. These values of n are all very much larger than the $n = 4$ chosen to illustrate the hard bubble structures in figure 4.3; this low value of n was only used to obtain a clear picture of the spin distrubution.

To put some numerical data into this problem, the value of the collapse diameter for normal bubbles in the bubble film under discussion was measured to be $D_{col} \approx 5.0\,\mu$m. For the case of the $n \approx 100$ bubble, this means that one complete cycle of M in (r, θ) occupied a distance of $\approx 0.1\,\mu$m along the circumference of the wall just before this particular bubble collapsed. This seems to be quite a reasonable result because $\Delta \approx 0.05\,\mu$m for this film and equation 4.18 tells us that the wall width at collapse is about one-third of the normal bubble wall width.

4.5 Hard Bubble Wall Mobility

We shall now consider the dynamic properties of hard bubbles and look first at the simplest of all possible dynamic experiments: the contraction or expansion of a bubble domain due to a change in the magnitude of the uniform bias field which is supporting it.

It is important to note that we consider really hard bubbles to begin with. These are bubbles of the kind described at the end of the last section with perhaps as many as 100 cycles in M as we go once around the bubble circumference. The dynamic properties of these very hard bubbles are more straightforward than the dynamics of so-called normal or soft bubbles because it is possible to assume that the dynamic wall structure is only a very small deviation away from the static wall structure. It will be recalled that this assumption was very important in chapter 2 where simple straight wall dynamics were considered.

Consider the bubbles shown in figure 4.3 in a state of uniform expansion. This means that the magnetic field at the wall must be below the value required to stabilise that particular bubble domain radius. For example, the simplest experimental arrangement would be where a pulsed magnetic field was arranged to subtract from the bias field which was initially holding the hard bubbles in equilibrium.

Concentrating upon the magnetisation at the centre of the wall region, which is represented by the small arrows in figure 4.3, the first effect of the drive field, B_z, is to produce a rotation of this magnetisation in (r, θ). In the case of the Landau–Lifshitz wall, considered in chapter 2, and in the case of the normal bubble, this rotation of M in the plane of the bubble film (figure 4.1) produced the canting of M out of the plane of the wall which was shown in chapter 2 to be the origin of the wall motion and also the origin of the effective wall mass. This really came about because the small component of M which developed normal to the wall plane, M_y, could interact with the large magnetic field in the plane of the wall, $\mu_0 M_x$. In the case of a normal bubble wall, M_y becomes M_r and $\mu_0 M_x$ becomes $\mu_0 M_\theta$, the field which circulates around the circumference of the normal bubble domain.

The hard bubbles, illustrated in figure 4.3, have no such circulating magnetic field because M_θ continuously alternates, as does M_r. The response of the wall magnetisations shown in figures 4.3a and b to a drive field B_z, which will be positive if it is to cause bubble expansion, will still be that the in-plane component of M will rotate in (r, θ) but this will simply cause the complex periodic pattern shown to circulate around the bubble wall, anticlockwise in the case of figure 4.3a and clockwise in the case of figure 4.3b, always assuming that we are dealing with a material which has γ negative.

This can be seen from the Landau–Lifshitz equation

$$\frac{-1}{|\gamma|} \frac{dM}{dt} = (M \times F) - \lambda[F - (F \cdot M)M/M_s^2] \qquad (4.28)$$

which was introduced in chapter 2 as equation 2.34. Because $(M \times F) = 0$ before any drive field is applied we may simply set $F_x = F_y = 0$, $F_z = B_z$ in equation 4.28 to find the initial response of the hard bubble to the pulsed increase in the bias field, B_z. It then follows from the x- and y-components of equation 4.28 that M rotates in (r, θ), at any point on the bubble wall, at an angular frequency $|\gamma|B_z$. The pattern of magnetisation shown around the wall in figures 4.3a and b rotates at the lower angular frequency $|\gamma|B_z/n$. This rotation gives us no reason to expect any change in the bubble size. The radial motion which does occur comes from the z-component of equation 4.28 which is simply

$$\frac{1}{|\gamma|} \frac{dM_z}{dt} = \lambda B_z \qquad (4.29)$$

At the wall centre, $r = R$, which may be identified with $y = 0$ in equation 4.13, we may write

$$\frac{dM_z}{dt} = \frac{\partial M_z}{\partial r} \frac{dr}{dt} \qquad (4.30)$$

and identify the radial expansion velocity, $v_r = dr/dt$, with dy/dt. An expression for this then follows from equations 4.30 and 4.29, using equations 4.13 and 4.18, and is

$$v_r = \frac{\alpha|\gamma|\varDelta}{[1 + (2n\varDelta/D)^2]^{1/2}} B_z \qquad (4.31)$$

when we use $\alpha = \lambda/M_s$ as we did in previous chapters.

Let us compare our result (equation 4.31) for the cylindrical hard bubble wall, (θ, z), with the simple Landau–Lifshitz planar wall, (x, z), result

$$v_y = (|\gamma|\varDelta/\alpha)B_z \qquad (4.32)$$

obtained in chapter 2 as equation 2.53. It is clear, from equations 4.31 and 4.32 that the mobility, μ_w, equation 4.1 of the hard bubble wall is at least α^2 times smaller than that of the simple Landau–Lifshitz wall or normal bubble wall. This represents a drastic reduction in wall mobility because $\alpha \approx 10^{-2}$ to 10^{-3} in good quality bubble domain garnet films.

There is, however, a rather subtle complication which makes this reduction in mobility less important than it would appear to be at first. In the simple Landau–Lifshitz model of chapter 2, the drive field must be kept well below the very small value $\mu_0\lambda/2$, that is $(\alpha/2)\mu_0 M_s$, for the model which leads to equation 4.32 to apply at all. In most practical cases this is such a small field that it is well below the coercive field of the material and, in fact, the low field mobility equation 4.32 is very rarely observed as we discussed at length in section 2.13. No such restriction applies to the drive field for the hard wall, however, because increasing the drive simply increases the rate of rotation of the magnetisation pattern around the wall.

Malozemoff and Slonczewski (1972) were able to measure both the normal (equation 4.32) and the hard (equation 4.31) wall mobilities in a sample of $(YbEuY)_3(FeGa)_5 O_{12}$ epitaxial garnet and found very good agreement with the model presented here. Similar agreement between experiment and theory was found by Vella-Coleiro *et al.* (1972b) but their data came from the bubble translation experiment which is the subject of the next section.

4.6 The Hard Bubble in Translation

In the bubble translation experiment (Vella-Coleiro and Tabor, 1972) a bubble domain is subjected to a uniform magnetic field gradient, pulsed on for a short period of time, typically between $0.1\,\mu s$ and $1.0\,\mu s$, this gradient being added to the constant bias field which is needed to support the bubble domain itself. From the simplest point of view, there should be a force acting upon the bubble domain which is in the same direction as the applied field gradient and we would expect to see the bubble move off in this direction. By observing the initial and final bubble positions and dividing the distance between these by the pulse length, a measure of the bubble translation velocity should be obtained if it is assumed that inertial effects are negligible. It was indicated in section 1.9, that this experiment turns out to be far more complicated than this simple point of view allows and, in the case of really hard bubbles the experimental facts are quite fascinating.

When a pulsed magnetic field gradient is applied to a really hard bubble it is observed to move virtually at right angles to the applied field gradient. What is even more remarkable is that two hard bubbles may be found to have almost indistinguishable static properties, that is they have identical diameter–bias field characteristics, and yet they will move in opposite directions when the pulsed magnetic field gradient is applied: one to the left of the gradient direction, the other to the right.

This interesting behaviour may be explained by asking what happens to the distribution of M within the bubble walls, shown in figure 4.3, when the gradient field is applied. The answer is shown in figure 4.4. The applied gradient field is such that the local drive field, B_z, is positive over the top half of the bubbles, as they are illustrated in figure 4.4, and negative over the bottom half. For the

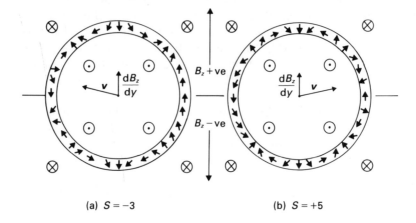

(a) $S = -3$ (b) $S = +5$

Figure 4.4 The wall structures shown in Figure 4.3 are both subjected to the same gradient in B_z which would normally be expected to translate these bubbles up the diagram. The hard wall structure is unwound on the right-hand side of the $S = -3$ bubble and is more tightly wound on the left-hand side. In the case of the $S = +5$ bubble, the opposite applies. The result is that the velocity, v, takes up an angle to the direction of the gradient, dB_z/dy

polarity of bubble shown, such a gradient field would normally drive the bubbles in the direction of dB_z/dy. Referring back to our equation of motion 4.28, and neglecting the damping terms in order to get just a qualitative picture, it is clear that the form of B_z described above will cause M to rotate anticlockwise in (r, θ) within the wall region of both top halves of the bubbles shown in figure 4.3, the initial state, while within the lower halves of both walls M will rotate clockwise.

This means that the distribution, or the pattern, of M will rotate anticlockwise within the top half of the wall of the $S = -3$ bubble, shown in figure 4.3a, but clockwise in the lower half of the wall. This will result in a new distribution of the kind shown in figure 4.4a where the hard bubble wall has become more tightly wound on the left-hand side but has unwound on the right. The opposite applies to the $S = +5$ bubble, which has the initial state shown in figure 4.3b. The effect of the gradient field on this bubble is shown in figure 4.4b.

Now we must remember that the gradient field, dB_z/dy, is added to the bias field, B_0, which was supporting this hard bubble in the first place. B_0 is negative, for the polarity of bubble domain shown, and will be well above the normal bubble collapse field, as can be seen from figure 4.2. If we consider the $S = -3$ bubble shown in figure 4.4a in the light of these remarks, it is clear what must happen. On the right-hand side of this bubble, the wall has become more like a normal bubble wall and the local value of the bias field will appear to be too large for equilibrium. This will mean that this part of the wall will move to the left. On the left-hand side of this $S = -3$ bubble the wall has become more tightly wound, or harder, than it was before. The local bias field will appear to be insufficient for equilibrium, this part of the bubble domain will try to expand and this, again, means wall motion to the left. The reason why the whole bubble should move to the left is now clear.

The opposite applies to the $S = +5$ bubble. As figure 4.4b shows, this bubble becomes more tightly wound on the right-hand side as a result of the applied gradient field and would therefore be expected to have a component of velocity to the right, as shown.

The hard bubbles will also have a component of velocity along the direction of the gradient because of the finite, although small wall mobility given by equation 4.31. This may be treated in a very simple way by going back to the constant mobility model equation 4.1 and using the well-known result

$$v_y = \mu_w (D/2)(dB_z/dy) \tag{4.33}$$

for the velocity of a bubble domain in a uniform gradient first given by Thiele (1971). Replacing μ_w in equation 4.33 by the mobility given by equation 4.31 gives

$$v_{//} = \frac{(\alpha|\gamma|D\varDelta)}{2[1 + (2n\varDelta/D)^2]^{1/2}} \frac{dB_z}{dy} \tag{4.34}$$

for the component of velocity parallel to the applied gradient. This simple argument neglects coercivity.

An estimate of the component of velocity perpendicular to the direction of the gradient may be made by saying that this will be equal to the angular velocity, at the leading or trailing parts of the wall, of the M pattern, $\pm|\gamma|(D/2)$ $(dB_z/dy)/n$, multiplied by the bubble radius, $D/2$. This is a very simple model but should be close to the truth because it implies that the hard bubble moves sideways at such a speed as to maintain the exchange energy of its wall at a minimum. The result is

$$v_\perp = \frac{|\gamma|D^2}{4n} \frac{dB_z}{dy} \tag{4.35}$$

Equations 4.34 and 4.35 may then be used to find the tangent of the hard bubble deflection angle as

$$\tan \beta = \frac{v_\perp}{v_{//}} = \frac{[1 + (2n\varDelta/D)^2]^{1/2}}{\alpha(2n\varDelta/D)} \tag{4.36}$$

which tends to α^{-1} when we remember that, by definition, a very hard bubble

must have $2n\Delta \gg D$. This last result was given by Slonczewski (1972b) and by
Thiele (1973b) who both included the effects of coercivity in their treatment
of the problem. The problem was treated in more depth by Thiele *et al.* (1973).

Any accurate treatment of the hard bubble translation problem would be
very complex because the pattern of M around the bubble wall can no longer
be assumed uniform and this will introduce a very complicated distribution of
damping forces. The result would certainly not be a linear one, as our approximations here (equations 4.34 and 4.35) have turned out to be. This is the case
even before the non-linearity of coercivity is included. The simple results which
have been given here lead to a fairly accurate picture for the really hard bubbles
under discussion.

The real difficulties of magnetic bubble domain dynamics stand out when
we leave the tightly wound structure of the really hard bubble and begin to deal
with a bubble having what we might call the hard wall structure over only part
of its circumference, the rest of its wall being normal. Before looking at this
problem, however, we have two more points to cover with really hard bubbles:
their generation and so-called suppression.

4.7 The Generation of Hard Bubbles

The hard bubble wall structure which was introduced in section 4.4 to explain
the static properties of hard bubbles and then used again in the previous two
sections to deal with their dynamics, has never been observed directly, by some
technique such as Lorentz microscopy or neutron scattering, and must be considered as only an hypothesis. There is, however, other experimental evidence
which makes us believe that the hard bubble wall structure described by equations
4.11 to 4.13 is very close to the truth. For example, Suzuki and Sugita (1978)
found bubbles in very thin garnet films which followed the normal bubble
diameter–field characteristic, like the one shown on the left of figure 4.2, and
then, at the normal collapse field and diameter, these bubbles changed diameter,
discontinuously, to one very well below D_{col}. This behaviour is well explained
by such bubbles having only a small segment of their wall belonging to the hard
bubble structure. When the diameter suddenly drops, it is only this part of the
wall which remains, now tightly wound. We shall be considering the theory of
these bubbles which have wall segments in section 4.9.

The strongest evidence giving further support to the hard bubble wall structure of equations 4.11 to 4.13 comes from a study of the generation of hard
bubbles. One of the earliest papers on hard bubbles (Tabor *et al.*, 1972b) described a method of hard bubble generation in some detail. A normal bubble
was subjected to a very large bias field pulse of a few microseconds' duration
which caused rapid bubble expansion. After the pulse was over, the normal
bubble would be seen to be replaced by a multifingered domain which would
become a hard bubble when the bias field was subsequently increased. This
multifingered appearance of domains in materials supporting hard magnetic

bubbles is shown in some excellent photographs included in the book on magnetic bubbles by Bobeck and Della Torre (1975), which also contains some interesting historical points about hard bubbles not published elsewhere.

While a large expanding pulsed field certainly will produce a hard bubble in many cases, the degree of hardness and the sign of S are both quite undefined. This would be expected because there is no reason to attribute any sense of left- or right-handedness to a simple uniform field and initial conditions having a cylindrical symmetry. Some kind of in-plane asymmetry must be introduced into the hard bubble generation procedure if a predetermined sign of S is to be obtained and a technique for doing this was worked out by Nishida *et al.* (1973). Their technique was to cut a stripe domain by means of pulsed currents in two conductors which passed over the stripe at two points about 150μm apart. The resulting short stripe domain could then be contracted into a bubble by increasing the bias field.

The details of these interesting experiments are shown in figure 4.5. It was found that the degree of bubble hardness and the sign of the bubble state number, which could be determined as shown in figure 4.5d by translating the bubble using the same conductors which had been used for stripe cutting, depended upon the angle between the stripe domain and the conductors during the cutting step, figure 4.5a, of the experiment. As figure 4.5d shows, the bubble deflection angle in translation always has the same sign as the cutting angle.

An explanation for these results may be given in terms of the hard bubble wall structure, discussed in the previous sections, and provides additional indirect evidence that this structure must be very close to the truth. In figure 4.6 we concentrate attention on what is happening on an atomic scale at the upper of the two conductors shown in figure 4.5 when the stripe domain is actually being cut by the pulsed current, I. For simplicity we shall assume that the conductor is transparent to the pulsed magnetic field so that the B_z field, due to the current I, has its maximum value close to the conductor edge. In figure 4.6 the conductor is shown passing over the stripe domain.

The field required to cut a stripe domain is above $\mu_0 M_s$ under static conditions and even greater when pulsed fields must be used (Kinsner *et al.*, 1974). Such high drive fields mean that we have a situation similar to the very high drive problem considered in sections 3.9 and 3.10, and the wall magnetisation will simply precess in the plane of the film as the wall moves forward. The drive field is non-uniform, however, so that different parts of the wall move at different speeds. As shown in figure 4.6, the part of the stripe domain wall which is moving fastest, at the edge of the conductor, must be the point where the magnetisation within the wall begins to wind up, very much like the hard bubble wall structure of figure 4.2.

Figure 4.6 shows how this winding up of M within the wall develops, $t = 0$, 1, 2, until the stripe domain is cut on the top edge of the conductor. If we now progress around the stripe head shown in the last frame, $t = 2$, of figure 4.6, using the S-number convention defined in section 4.4, we find that M makes

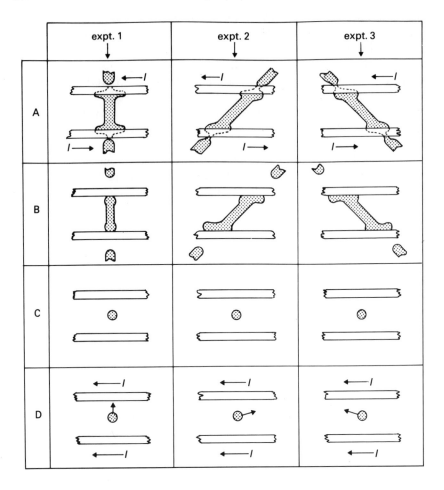

Figure 4.5 Showing the experiments of Nishida *et al.* (1973) which gave controlled generation of hard magnetic bubbles. In each experiment, a stripe is cut by passing opposing pulsed currents in conductors laid over the stripe, A, to leave an isolated domain, B. This is contracted into a bubble, C, which is then examined in translation, D, using aiding pulsed currents in the two conductors to provide the required gradient field

−3, +2, −2, +3 rotations in the plane of the film as we progress completely around the stripe head and the resulting stripe, and thus the bubble domain it forms when the bias field is increased will be normal because these sections of wall with opposite winding senses will unwind into one another.

When the stripe domain is inclined at a positive angle to the conductors, as shown in figure 4.5, experiment 2, the winding numbers given in the previous paragraph will be modified so that they no longer add up to zero. As shown in

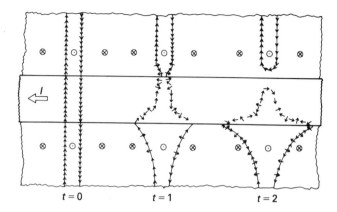

Figure 4.6 **Showing the winding up of the wall structure during the stripe cutting process**

figure 4.5a, far greater motion takes place in experiment 2 on the left-hand side under the top conductor and on the right-hand side just outside the top conductor. As explained previously, these regions are where positive winding takes place during the cut and the resulting bubble should have a positive S-number and be deflected to the right when tested in translations. This was confirmed experimentally by Nishida *et al.* (1973), as shown in figure 4.5d, experiment 2.

The opposite argument applies when the stripe domain makes a negative angle with respect to the cutting conductors, as shown in figure 4.5, experiment 3. The stripe now expands predominantly where negative winding of the wall structure occurs and a bubble with a negative S-number results.

These elegant and simple experiments by Nishida *et al.* (1973) clearly support the hard bubble wall model which has been presented here because they give a quantitative reason behind the sign of the hard bubble deflection and also the degree of hardness. Nishida *et al.* (1973) found a clear connection between the magnitude of the hard bubble collapse field and the tilting angle.

4.8 The Suppression of Hard Bubbles

The dynamic properties of hard magnetic bubbles are clearly undesirable for bubble domain devices and the early papers included some discussion of empirically determined methods which had been found to apparently suppress the hard bubble phenomena. These methods involved the growth of multilayer epitaxial garnet films (Bobeck *et al.*, 1972a 1972b) or ion-implantation (Wolfe and North, 1972, Wolfe *et al.*, 1973, North *et al.*, 1978). It was originally thought that the domain wall which would appear at the top of the bubble domain in either a multilayer or an ion-implanted film, that is the wall in between the two layers which would form some kind of cap over the top of the bubble, would cause the tightly wound hard bubble wall structure to unwind (Rosencwaig, 1972).

Later work by Henry *et al.* (1973), however, showed that this was not so and that the existence of hard bubbles was not prevented by multilayering or by ion-implantation. It was the generation of hard bubbles which these techniques prevented and, to be more exact, it was the temperature below which hard bubble behaviour could be observed that the hard bubble suppression process had reduced. For example, an ion-implanted film which showed no hard bubble behaviour at room temperature would have hard bubbles if it was cooled. Furthermore, hard bubbles which were generated at the low temperature would continue to exist in the film when it was heated up to room temperature again.

These findings were soon confirmed by other workers who discovered special garnet compositions which did not show hard bubble behaviour (Smith *et al.*, 1973, Smith and Thiele, 1973), eliminated hard bubbles by depositing a thin layer of permalloy on top of the garnet film (Takahashi *et al.*, (1973a) or found that hard bubbles could not be generated in a film if it was grown epitaxially on a slightly misaligned substrate (Hoshikawa *et al.*, 1974). All these techniques only reduced the temperature below which hard bubbles could be generated in these films and, once generated, these hard bubbles would survive if the film was heated up.

It is not surprising that the existence of hard bubbles is unaffected by a modification to the surface properties of the bubble domain film. Only a very small modification is needed to suppress hard bubble generation; Takahashi *et al.* (1973a) found that only 50 Å of permalloy deposited on top of the garnet film was sufficient while Lin and Keefe (1973) found 200 Å was enough but added the important observation that a layer of only 100 Å of SiO_2 in between the garnet layer and the permalloy film would completely destroy the hard bubble suppression effect. This was confirmed by Takahashi *et al.* (1973b) and by Suzuki *et al.* (1975).

Armed with all these experimental findings it is possible to suggest the mechanism whereby hard bubble generation may be suppressed in bubble domain films. The work with special garnet compositions which were free from hard bubbles (Smith *et al.*, 1973, Smith and Thiele, 1973, Hoshikawa *et al.*, 1976) and with films grown on misaligned substrates (Hoshikawa *et al.*, 1974) clearly showed that an in-plane anisotropy was necessary to prevent hard-bubble generation. The same must apply for multilayered, ion-implanted and permalloy coated bubble garnet films but in these cases the in-plane magnetic anisotropy must come from exchange coupling to the in-plane magnetisation of the capping layer. All the capping layers have in-plane magnetisation, the ion-implanted layer having this because the strain caused by ion-implantation overcomes the growth-induced magnetic anisotropy of the main garnet layer (North *et al.*, 1978). The fact that the coupling between the in-plane direction of M in the capping layer and the main layer must be through exchange, and not simply magnetostatic, is suggested by the experiments described above which involved the thin layer of SiO_2 in between the garnet film and the permalloy capping layer. The fact that the magnetisation of the capping layer does, in fact, lie in-plane may be deduced from the domains which are seen when the surface is decorated with magnetic colloid. Both permalloy capping layers (Suzuki *et al.*, 1975) and ion-implanted garnets (Puchalska *et al.*, 1977) have been decorated in this way. Some further information about what is going on inside the capping layer may be obtained from FMR measurements (Soohoo, 1978).

Hard bubble generation must be suppressed when there is an in-plane preferred direction in a bubble film because bubble expansion under a pulsed bias field, which would normally produce hard bubbles, will then produce a symmetric form of expanding bubble, the kind of domain discussed in connection with figure 4.5, experiment 1. This is conjecture and, clearly, much more experimental work needs to be done in this area before we can say that this problem is fully understood. High speed microphotography is certainly the most promising line of approach and Gál *et al.* (1975) have already published photographs showing the kind of shapes which bubble domains take, in both ion-implanted and as-grown films, when subjected to intense expanding bias field pulses. These photographs show that the expanding bubbles turn into very complex shapes but have a regular symmetry in the ion-implanted films and a random shape in the as-grown films. Kleparskii *et al.* (1977) have also looked at this kind of problem with high speed microphotography and have used an improved technique compared to the conventional low repetition rate dye laser (Humphrey, 1975), which is really a sampling technique suited only to repetitive events which hard bubble generation certainly is not. Kleparskii *et al* (1977) used a solid state laser and an image intensifier which enabled them to take genuine multiple exposure photographs of one particular bubble event. High-speed photographs which actually show the details of hard bubble generation, and might answer some of the questions raised here, have yet to be taken, however. It may well turn out that our model of the hard magnetic bubble, after all, requires some serious modifications.

4.9 The Statics of Bubble Domains with Low S-numbers

If the models of hard bubble generation and the suppression of hard bubble generation, given in the previous two sections, are correct it is obvious that an exact cancellation of opposite sense winding around the wall of a domain is not very likely and what we have been calling a normal bubble, illustrated ideally in figure 4.1, may contain a few Néel lines. In this section we consider the static properties of such domains. The problem is also the same one as asking what happens to a hard bubble when the bias field is reduced so that the hard wall structure is no longer tightly wound.

The kind of situation which should arise is shown in figure 4.7. According to our previous definition of S-number, given in section 4.4, the bubble domain shown in figure 4.7 has $S = +6$. The alternations of M in (r, θ) do not occur uniformly around the circumference of the bubble, however, but are now crowded into a segment of the wall which subtends the angle ψ. The rest of the wall shown in figure 4.7 is normal.

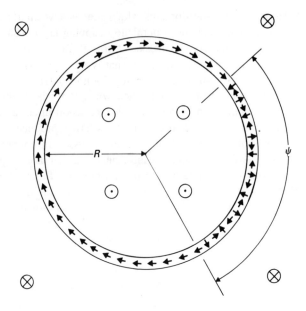

Figure 4.7 When the number of alternations of M_θ in a bubble domain is small,
these will be packed into a small segment of the wall because magnetostatic forces will make the normal part of the wall expand

This asymmetric distribution in the winding of M is a lower energy configuration compared to a uniform distribution because the magnetostatic energy of
the normal part of the wall is reduced if this part of the wall expands and com-
presses the hard wall section, thus increasing its exchange energy, until a balance
is found. This effect has been considered, from a rather different point of view,
by Hubert (1973) and Slonczewski (1974a) and termed 'vertical Bloch line
clustering'.

The value of ψ which minimises the total energy of the bubble domain wall,
shown in figure 4.7, must now be found. For the hard wall section, we shall
only consider exchange and anisotropy and these contribute an energy density
given by equation 4.16, which may be simplified in this case by writing
$\delta y = \Delta$, from equation 4.18, because the tight winding condition is now relaxed,
and setting $\delta x = \psi D / 4n$, which clearly follows from figure 4.7. The total energy
of the hard wall section is then found by integrating with respect to y, again
using the limits $-\infty$ to $+\infty$, and multiplying by the area of the hard wall segment
$\psi Dh/2$. The result is

$$E_{hw} = 2A\psi Dh\Delta(8n^2\pi^2/\psi^2 D^2 + 1/\Delta^2) \qquad (4.37)$$

The normal wall section has the well known surface energy density $4(AK_u)^{1/2}$
due to exchange and anisotropy (Brown, 1962); this term occurred in equation

4.8, and also has a magnetostatic energy density due to the magnetic field $\mu_0 M_x$ along its length. Using $M_x = M_s$ sech $(y/\varDelta)$ again, integrating and multiplying by the area of the normal wall section $(2\pi - \psi)Dh/2$, gives the result

$$E_{nw} = \tfrac{1}{2}(2\pi - \psi)Dh(4A/\varDelta - \mu_0 M_s^2 \varDelta) \qquad (4.38)$$

for the total energy.

Adding equations 4.37 and 4.38, to give the total energy of the entire wall, shows that only two terms remain in the result which depend upon ψ. As we might expect, these are the magnetostatic energy of the normal wall section, which gets smaller the smaller ψ becomes, and the additional exchange energy due to the winding up of the hard wall section, which gets larger the smaller ψ becomes. The sum of these two terms is a minimum when $\psi = \psi_0$ where

$$\psi_0^2 = (32\pi^2 n^2 A)/(\mu_0 M_s^2 D^2) \qquad (4.39)$$

Using the definition of the material characteristic length, l, given by equation 4.8, and the definitions of Q and $\varDelta$ makes it possible to write this result very concisely as

$$\psi_0/2\pi = (n/Q^{1/2})(l/D) \qquad (4.40)$$

This problem was first considered by Hubert (1973) who showed that the equilibrium spacing between Néel lines, or vertical Bloch lines, should be of the order of $\pi l/(2Q)^{1/2}$, written $(\pi A)^{1/2}/M_s$ in the old system of units. This result came from modelling the closed finite bubble wall by an infinite planar wall, so that the magnetostatic force coming from the normal wall section which has been introduced here was omitted. The reason for the clustering of Néel lines in Hubert's model is the magnetostatic attraction between the segments of the hard wall structure themselves, a magnetostatic interaction which has been omitted here because it is considered to be negligible compared to the stronger magnetostatic term coming from the normal part of the wall. This conclusion is supported by the fact that the result given here, equation 4.40, predicts a smaller equilibrium spacing between Néel lines, $\pi l/2Q^{1/2}$, compared to Hubert's result $\pi l/(2Q)^{1/2}$. The difference is, however, not very important in view of the approximations involved in both models. The important point is that the structure of the low S-number bubbles must have the asymmetric form shown in figure 4.7. In the case actually shown in figure 4.7, where $n = 5$, the hard wall section would occupy a very small segment of the total wall. The wall width in figure 4.7 has been exaggerated again for clarity.

Experimental evidence for the fact that bubble domains and stripe domains will often contain hard wall sections, of the kind being considered here, comes from high-speed microphotography. Morris *et al.* (1976) have given some very clear photographs of an expanding stripe domain which showed how short segments of the domain wall remained almost immobile while surrounding parts of the stripe underwent rapid expansion. A series of photographs was used to compare the sideways motion of the hard wall section and its forward mobility and the result was in fair agreement with theory. Craik and Myers (1975) also observed stripe domains having different wall mobilities at different points in one and the same sample of material. These authors used a more indirect magneto-optical technique.

Additional evidence in support of hard wall sections is the observation of short dumb-bell domains in static equilibrium with an externally applied bias field which also supports normal and hard bubble domains in the same film. Elliptical domains are also observed in these hard bubble materials. As the bias field is increased, the dumb-bell and elliptical domains may be observed to convert into hard bubbles or to collapse as dumb-bells or ellipses at bias fields very close to normal collapse. Experiments which give some detailed data on this interesting behaviour have been described by Malozemoff (1972) and by Slonczewski *et al.* (1972).

Figures 4.8 and 4.9 show the detailed wall structure which may explain this kind of behaviour. In figure 4.8 two elliptical domains are shown in which the magnetisation circulates in a clockwise direction within most of the wall region but there are two short hard wall sections. The hard wall sections are shown

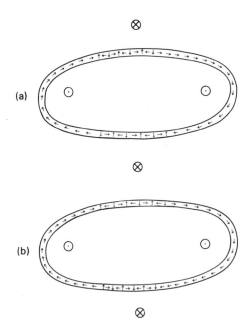

Figure 4.8 Two possible structures which would give an elliptically shaped domain in a bubble domain film at a bias field close to the bubble run-out field

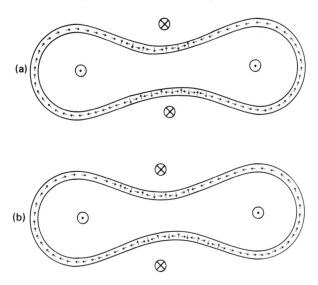

Figure 4.9 Two possible structures which would give a dumb-bell shaped domain in a bubble domain film at a bias field close to the bubble run-out field

occupying the regions of lower curvature in the wall because of the higher surface tension of the hard wall structure. The same applies to the two dumb-bell domains shown in figure 4.9—again the hard wall sections avoid regions of high curvature. The difference between the elliptical and dumb-bell domains is in the magnetostatics of the normal wall sections. In the elliptical domain, the two normal wall sections attract one another and this helps to hold the domain together. In the case of the dumb-bell domain the interaction between the normal wall sections is repulsive so that these wall sections try to close upon themselves to reduce the magnetostatic energy and the dumb-bell shape results.

The difference between the two domains shown in both figures 4.8 and 4.9 is in their state numbers. Counting the number of revolutions M makes in the plane of the film as we travel completely around these domains once, it is clear that the domain shown in figure 4.8a has $S = 1$ while for figure 4.8b, $S = -5$. For figure 4.9a $S = +2$ while for figure 4.9b $S = -7$. These state numbers are very much lower than would be found in practice for elliptical and dumb-bell domains; as before, the wall width has been exaggerated in the diagrams to give a clear illustration.

As the bias field is increased on an elliptical or a dumb-bell domain, the two hard wall sections must eventually be forced together and will annihilate one another if they have opposite winding numbers, as in figures 4.8a and 4.9a, to leave a normal bubble, or a very close to normal bubble, which may well find itself above normal collapse conditions and vanish before any observation of its existence may be made. If the hard wall sections are wound in the same sense, as they are in figures 4.8b and 4.9b, increasing the bias field can produce a hard bubble from the elliptical or dumb-bell shaped domain.

The dumb-bell domains have very interesting dynamic properties, first discussed by Slonczewski *et al.* (1972) and these were apparently the first problem in magnetic bubble domain dynamics to be studied using high speed microphotography (Slonczewski *et al.*, 1972). Interesting though these properties may be—the dumb-bells are found to rotate in a pulsed bias field and the sense of rotation depends upon the sign of S—the elliptical and dumb-bell domains are very special and rather rare entities and we should return to the low S-number bubble domains which are now our main problem for the rest of this chapter.

Really low S-number bubbles are illustrated in figure 4.10. Here, we have again been forced to exaggerate the angle which is subtended by the hard wall section so that the in-plane alternations of M may be clearly seen. There are two possible forms for each S-number and this comes from the screw-sense of the

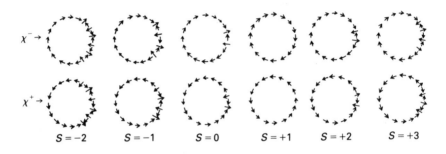

χ^- →

χ^+ →

| $S=-2$ | $S=-1$ | $S=0$ | $S=+1$ | $S=+2$ | $S=+3$ |

Figure 4.10 **Illustrating the possible structures of bubbles with small S-numbers**

normal wall section. Following Argyle *et al.* (1976c) we label these two possibilities χ^+ and χ^- the sign $^+$ or $^-$ being the sense of rotation of M_θ within the normal wall section relative to the sign of M_z inside the bubble. The usual right-hand screw rule is used to determine the rotation sense, relative to M_z inside the bubble domain.

4.10 The Dynamics of Bubble Domains with Low S-numbers in Small Applied Fields

We now come to the main problem of this chapter: the dynamics of the low S-number bubbles, the kind illustrated in figure 4.10. These are the normal or soft bubbles used in the bubble domain device.

In this section, we shall restrict the discussion to small drive fields and translational motion in a simple gradient field. Section 4.13 of this chapter considers the problem of bubble translation under conditions of high drive.

By definition, a small drive field is one which may be assumed to produce no dramatic change in the wall structure of the moving bubble, relative to its static structure. This problem was discussed quantitatively at the beginning of section 3.2, and we begin here by looking at the experimental problems of bubble translation to see how well the small drive field requirement might be satisfied in practice.

4.10.1 Experimental Problems

In the bubble translation experiment, described in section 1.9, an isolated bubble domain, constant diameter D, is subjected to a simple field gradient dB_z/dy. If the constant wall mobility model

$$v_n = \mu_w(B_z - B_c) \qquad (4.41)$$

is applied to this problem the well known result (O'Dell, 1974) for the bubble translation velocity

$$v_y = \mu_w[(D/2)(dB_z/dy) - (4/\pi)B_c] \qquad (4.42)$$

is obtained where B_c is a material constant termed the coercive field. This result was first stated by Perneski (1969).

Many of the difficulties of the bubble translation experiment were discussed in chapter 1. There is the problem of correct bias compensation, that is attempting to maintain D constant, and the problem that the simple constant mobility model equation 4.41 is clearly quite inapplicable to this experiment in that the bubble domain rarely moves in the direction of the gradient anyway. Nevertheless equation 4.42 is a useful expression for these introductory remarks when we wish to consider two very important problems connected with bubble translation. These are coercivity and what we may expect to be a low drive field from a quantitative point of view.

To consider coercivity first, there has always been an interesting difference in the coercivity of a bubble domain device film when it is measured by using equation 4.42 and when it is measured by simple hysteretic techniques involving the sinusoidal excitation of a stripe domain pattern. The coercivity measured by the latter method is very much lower, as a rule, than the coercivity measured by bubble translation, that is by finding the value of (dB_z/dy) which will just cause motion of a bubble and using equation 4.42 to write

$$B_c = (\pi D/8)(dB_z/dy)_{v_y \approx 0} \qquad (4.43)$$

Once the experimental difficulties are understood, and these have been dealt with in an excellent paper by Clover *et al.* (1972), very small fields are found to be needed to just start the motion of a straight domain wall in a magnetic bubble film. Clover *et al.* (1972) measured a straight wall coercivity of only $4\,\mu T$ in three different films of $(GdYYb)_3(FeGa)_5O_{12}$ grown from BaO fluxes. Walling (1979) made a very thorough study of a whole range of calcium germanium garnet films, intended for bubble domain devices using bubble diameters between $4\,\mu m$ and $1.5\,\mu m$, and obtained straight wall coercivities between $31\,\mu T$ and a record low of $1.4\,\mu T$ for the thinnest of these films, intended for $1.5\,\mu m$ bubble device work. Such low coercivities would be expected in crystals of the high perfection which liquid phase epitaxial garnets should have.

Bubble domain translational coercivities are, by contrast, very much higher and the reason for this is simply that the drive field around a bubble in translation is not constant, as it is along the straight wall in the stripe domain measurement. If the bias compensation is perfect, the two diametrically opposed points on either side of the translating bubble domain see no drive field at all and motion here can only take place by distortion of the bubble domain causing a local unbalance between the magnetostatic field and the bias field. This problem is avoided in the simple analysis which leads to equation 4.42 because the component of velocity normal to the bubble wall is zero at the sides of an ideally rigid bubble domain which is undergoing translation in the direction of the applied gradient. A detailed analysis of this coercivity problem has been given by Walling (1979) who showed that the coercivity which would be deduced from bubble translation, equation 4.43, could be over ten times the coercivity measured from straight wall motion. Walling (1979) obtained good agreement between his theory and measurements on the calcium germanium films mentioned above. Values of B_c obtained from equation 4.43 in these films, were between $43\,\mu T$ and $110\,\mu T$. Similar results had been obtained by Fontana and Bullock (1976), who found a translational coercivity of $30\,\mu T$ at room temperature in a $(YSmLuCa)_3(FeGe)_5O_{12}$ film and noted that this dropped dramatically when the film was heated up to $70\,^{\circ}C$. Clover *et al.* (1972) found translational coercivities of 31, 18 and $24\,\mu T$ in the three films referred to above which all had straight wall coercivities of only $4\,\mu T$.

This discussion of coercivity leads us back to the second question concerning our experimental problems which was raised at the beginning of this section: the quantitative conditions which would constitute low drive. The answer follows from the values of translational coercivity quoted above and equation 4.43. For example, if we take $30\,\mu T$ as an optimistic value of translational coercivity for a $5\,\mu m$ bubble, equation 4.43 tells us that the minimum (dB_z/dy) which will just cause bubble movement is $15.3\,T/m$. In order to collect any experimental data we may argue that the gradient field used in bubble translation would have to be increased to at least twice this figure so that we would be using $dB_z/dy = 30.6\,T/m$. The maximum drive field at the bubble wall, $(D/2)$ (dB_z/dy) would now be $\approx 80\,\mu T$ for a $5\,\mu m$ diameter bubble. If we turn back to section 3.2, where some numerical data was given on what would constitute low drive conditions in a bubble domain film with $\mu_0 M_s = 20\,mT$ and $\alpha = 10^{-4}$

it will be seen that the drive field was to be kept below $1.0\,\mu T$, that is below $\alpha\mu_0 M_s/2$. Clearly, our $80\,\mu T$ drive field is unacceptably high.

Two important conclusions can be drawn. The first is that low drive conditions may be arranged in a bubble film provided the damping is high, that is α is much larger than the 10^{-4} quoted above which is a value more typical of pure YIG. The substituted YIG used for bubble work may have a value of α as low as 10^{-2} (Hansen, 1974). This would put the critical drive field, $\alpha\mu_0 M_s/2$, quoted above up to $100\,\mu T$ for the material with $\mu_0 M_s = 20\,mT$ and then the $80\,\mu T$ drive field might be considered just acceptable.

The second conclusion we may draw is that more attention should be paid to experiments using very small bubbles. Micron-sized bubble materials have higher values of $\mu_0 M_s$ compared to the earlier $5\,\mu m$ bubble diameter materials and consequently have higher values of critical field, $\alpha\mu_0 M_s/2$, for the same value of α. What is vitally important, however, is that the most reliable experimental data shows that coercivity changes very little as bubble size is reduced (Vella-Coleiro *et al.*, 1979) and, in fact, Walling (1979) found that the straight wall coercivity went down as bubble size was reduced. Using the data given by Walling (1979) for a $1.5\,\mu m$ bubble film with $\mu_0 M_s = 47\,mT$ and his worst case translational coercivity of $43\,\mu T$, equation 4.43 gives the maximum of B_z to be $110\,\mu T$, when we use twice the coercivity limited drive gradient, whereas $\alpha\mu_0 M_s/2$ is $235\,\mu T$ when we take $\alpha = 10^{-2}$.

Clearly, the smaller sized bubble is easier to fit into the experimental constraints and this is fortunate because we shall see that one of the experimental variables we wish to measure, the bubble deflection angle, is also more accurately determined when we work with smaller bubbles.

4.10.2 The Bubble Deflection Angle

The sideways force which appears to deflect a low S-number bubble away from the expected direction of motion, in the simple bubble translation experiment, was discussed briefly in section 1.9, in connection with figure 1.13. This deflection of soft bubbles in translation might be thought at first to be a simple extension of the reasoning given in section 4.6 to explain the very large deflection angles which are observed with hard bubbles and an argument along these lines would lead us to expect that the $S = +1$ bubble, shown in figure 4.11, would be the bubble domain which would travel along the direction of the gradient while bubbles with $S = 0$ or negative S-numbers would be deflected to the left, as in figure 4.4a and bubbles with $S > +1$ would be deflected to the right.

The discussion in section 1.9 in connection with figure 1.14, showed that there is experimental evidence to suggest that the $S = +1$ bubble domain does have a deflection effect in its own right. The first attempt to give a clear explanation for this, was made by Thiele (1973b) who considered the problem from the following point of view: if we define the structure of the bubble domain and its wall, that is, we define the vector field M, and we define the velocity at

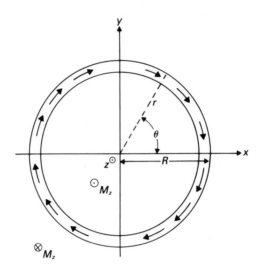

**Figure 4.11 The coordinate system used with the $S = +1$ bubble domain and its
M field**

which this vector field is moving, thus defining the field $(\mathrm{d}M/\mathrm{d}t)$, it should be
possible to take our equation of motion

$$\frac{-1}{|\gamma|} \frac{\mathrm{d}M}{\mathrm{d}t} = (M \times F) - \lambda[F - (F \cdot M)M/M_s^2] \qquad (4.44)$$

which was introduced in chapter 2 as equation 2.34; substitute the functions
describing M and $\mathrm{d}M/\mathrm{d}t$ and then solve equation 4.44 for the magnetic field F
which is needed to support the hypothetical structure and its motion. This is
obviously an approach which depends on some very serious assumptions about
the dynamic wall structure, particularly in view of the remarks which were made
in the previous section concerning the conditions which are likely to apply in
the actual experiment. Nevertheless, Thiele (1973b) obtained some very
interesting results and an interpretation of his method is given below.

Following Thiele (1973b), we write equation 4.44 using suffix notation

$$\frac{-1}{|\gamma|} \frac{\mathrm{d}M^k}{\mathrm{d}t} = e^{ijk}M_iF_j - \lambda(F^k - (F^iM_i)M^k/M_s^2] \qquad (4.45)$$

because it is then very easy to manipulate the equation into a more convenient
or useful form. In equation 4.45, the suffices take the values 1, 2, 3 for the three
space coordinates and when these are the Cartesian coordinates x, y, z, as they
will be in what follows, there is no distinction between upper and lower suffices.
The component M_z, for example may be written M_3 or M^3. Upper and lower

suffices are maintained in the notation, however, to allow us to use the rule that summation takes place over repeated suffices only when these occur as an upper-lower pair. Upper and lower suffices are then helpful in this kind of algebra because we know that our equations must always balance for both kinds of suffix. A useful reference for this notation is the book by Sokolnikoff (1951).

Equation 4.45 may be written

$$\frac{-1}{|\gamma|} \frac{dM^k}{dt} = e^{ijk} M_i F_j - (\lambda/M_s^2)[F^k(M^iM_i) - M^k(F^iM_i)] \qquad (4.46)$$

so that we may take out M_i as a factor to leave $(F^kM^i - M^kF^i)$ behind in the last term which may then be written using the generalised Kronecker delta, that is

$$\frac{-1}{|\gamma|} \frac{dM^k}{dt} = e^{ijk} M_i F_j - (\lambda/M_s^2) M_i \delta^{ki}_{lm} F^l M^m \qquad (4.47)$$

We now make use of the identities

$$\delta^{ki}_{lm} \equiv \delta^{kij}_{lmj} \equiv e^{kij} e_{lmj} \qquad (4.48)$$

to write equation 4.47 as

$$\frac{-1}{|\gamma|} \frac{dM^k}{dt} = e^{ijk}[M_i F_j - (\lambda/M_s^2) e_{lmj} F^l M^m M_i] \qquad (4.49)$$

and, going back to equation 4.46, write

$$-e_{lmj} M^m F^l = \frac{-1}{|\gamma|} \frac{dM_j}{dt} + (\lambda/M_s^2)[F_j(M^iM_i) - M_j(F^iM_i)] \qquad (4.50)$$

by making $(i\ j\ k)$ into $(l\ m\ j)$ in equation 4.46, interchanging lower and upper suffices and rearranging. Equation 4.50 may then be substituted into the last term of equation 4.49 to give the very simple result

$$\frac{-1}{|\gamma|} \frac{dM^k}{dt} = e^{ijk} M_i \left[F_j - (\alpha/|\gamma|M_s) \frac{dM_j}{dt} \right] \qquad (4.51)$$

when we write $\alpha = \lambda/M_s$, as usual, and neglect the term involving α^2.

Equation 4.51 is the same one which was referred to in chapter 2, equation 2.36, and attributed to Gilbert and Kelly (1955). Thiele (1973b) argued that equation 4.51 expresses a balance between the time rate of change of angular momentum on the left-hand side and the torque due to reversible effects plus the torque due to dissipative effects on the right-hand side.

We now apply this equation of motion to the problem of the $S = +1$ bubble domain in translation. It is clear that the component of the applied field gradient which overcomes the dissipative effects must be parallel to the velocity because work must be done. If there is any component of the field gradient which is perpendicular to the velocity, that is, if there is any deflection of the bubble, the origin of this must be in first two terms of equation 4.51 which make up the undamped equation of motion

$$\frac{-1}{|\gamma|} \frac{dM^k}{dt} = e^{ijk} M_i F_j \qquad (4.52)$$

Now let M^k be a function of $(x^m - v^m t)$ where x^m is the laboratory frame of reference and v^m is a velocity vector of constant magnitude and direction. In a frame of reference which moves with the bubble domain we then have

$$\frac{dM^k}{dt} = -v^m \frac{\partial M^k}{\partial x^m} \qquad (4.53)$$

and equation 4.52 becomes simply

$$v^m \frac{\partial M^k}{\partial x^m} = |\gamma| e^{ijk} M_i F_j \qquad (4.54)$$

Now let us choose the y-direction as the direction of motion and, for clarity write equation 4.54 as the matrix equation

$$(v_y/|\gamma|) \begin{pmatrix} \partial M_x/\partial y \\ \partial M_y/\partial y \\ \partial M_z/\partial y \end{pmatrix} = \begin{bmatrix} 0 & -M_z & M_y \\ M_z & 0 & -M_x \\ -M_y & M_x & 0 \end{bmatrix} \begin{pmatrix} F_x \\ F_y \\ F_z \end{pmatrix} \qquad (4.55)$$

where we now specify (x, y, z) as our coordinate system (x^1, x^2, x^3), figure 4.11. Equations 4.55 do not have a solution in general because the matrix involving the components of M is singular. To proceed we must consider a particular M field and choose the $S = +1$ bubble which was shown originally in figure 4.1 and more recently in figure 4.10. This may be represented approximately by

$$M_z = -M_s \tanh[(r-R)/\varDelta] \tag{4.56}$$

$$M_x = -cM_s \sin\theta \operatorname{sech}[(r-R)/\varDelta] \tag{4.57}$$

$$M_y = +cM_s \cos\theta \operatorname{sech}[(r-R)/\varDelta] \tag{4.58}$$

when the z-dependence of the wall structure is neglected. The variables r and θ are defined in figure 4.11 and c represents the chirality of the bubble domain, that is, $c = +1$ if M_θ circulates anticlockwise within the wall and $c = -1$ for a clockwise rotation of the kind shown in figure 4.11.

Having defined the structure of the moving bubble we can now consider the centre of the wall, $r = R$, and the central plane of the bubble film, $z = 0$. Equation 4.55 now gives a solution for F_z in terms of our defined M and this is simply

$$F_z = \frac{(v_y/|\gamma|)}{M_y}\frac{\partial M_x}{\partial y} = \frac{-(v_y/|\gamma|)}{M_x}\frac{\partial M_y}{\partial y} \tag{4.59}$$

In the coordinate system of figure 4.11, $x = r\cos\theta$, $y = r\sin\theta$, we have

$$\frac{\partial M}{\partial y} = \sin\theta\,\frac{\partial M}{\partial r} + \frac{\cos\theta}{r}\frac{\partial M}{\partial\theta} \tag{4.60}$$

and, substituting equations 4.57 and 4.58 into either form of equation 4.59, using equation 4.60 and setting $r = R$ in the final result, gives

$$F_z = -(v_y\cos\theta)/|\gamma|R \tag{4.61}$$

Note that the chirality parameter, c, cancels out and our solution does not depend upon the chirality of the bubble.

The field F_z can only come from an external source and it is clear from figure 4.11 that a field B_z which varies linearly with x will provide a field which varies as equation 4.61 if we have

$$\partial B_z/\partial x = -v_y/|\gamma|R^2 \tag{4.62}$$

Equation 4.62, then, gives the component of the applied gradient which must lie at right angles to the velocity vector of a translating $S = +1$ bubble.

To find the component of the applied gradient which must be parallel to the velocity vector, we may use the simple constant mobility model equation 4.41 and its consequence equation 4.42. This gives us

$$\partial B_z/\partial y = (v_y/\mu_w R) + (4B_c/\pi R) \tag{4.63}$$

The sign of the x-component of the applied field gradient, equation 4.62, depends only upon the sign of v_y and the sign of $|\gamma|$, tacitly assumed to be negative and always written $-|\gamma|$. The sign of the y-component, equation 4.63, also depends upon the sign of v_y but it also depends upon the sign of M within the bubble which we have always taken to be positive.

The situation is summarised in figure 4.12 where the predominant deflection angle of the $S = +1$ bubble domain in translation is shown. This sign of deflection angle and its relation to the direction of the bias field have been confirmed in a number of laboratories for bubbles in ion-implanted garnet films which would be expected to have the normal, or $S = +1$, wall structure for reasons which have been given in section 4.8.

4.10.3 The Physical Origin of the S = +1 Bubble Deflection

The deflection of the $S = +1$ bubble is, perhaps, an unexpected result when we look back at section 4.6 and compare the work done there on the deflection of hard magnetic bubbles. In the case of the hard bubble, its sideways motion came about because of the clear asymmetry which developed in its complex wall structure, figure 4.4, as a result of applying the drive field. Looking now at figure 4.10, the low S-number bubbles like $S = -2, -1, 0, +2$ or $+3$ also have a complex wall structure which we might expect to give some kind of deflection in translational motion. Why, however, should the perfectly symmetrical $S = +1$ bubble show any deflection?

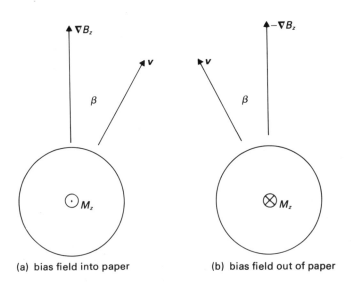

(a) bias field into paper   (b) bias field out of paper

Figure 4.12 **The relationship between the sign of the $S = +1$ bubble deflection angle and the sign of the bias field**

The argument given in the previous section leading to equations 4.62 and 4.63 looks formally acceptable but, if it is correct, there must be some simple explanation which gives a convincing physical origin for this $S = +1$ bubble deflection. It is interesting to note that none of the early papers (Slonczewski et al., 1972, Slonczewski, 1972b, 1974a, and Thiele, 1973b, 1974) give any explanation and later review papers (North et al., 1978, Slonczewski and Malozemoff, 1978) refer back to earlier or unpublished work.

This is a very unsatisfactory state of affairs and puts this part of the theory of magnetic bubble domain dynamics in question. The model of the bubble structure which was used (equations 4.56, 4.57 and 4.58) in the previous section to obtain the results in equations 4.62 and 4.63 is so crude that the same expressions should follow from a much more straightforward argument.

Let us consider a magnetic bubble moving along y at constant velocity, v_y, as shown in figure 4.13a. We suppose that M_z is positive within the bubble domain, as in all previous work. If we consider a small segment of the wall, length $R\,d\theta$ and width Δ, M_z must change from being $-M_s$ outside the bubble to $+M_s$ inside the bubble as this small region passes. This is true no matter what the internal structure of the wall may be.

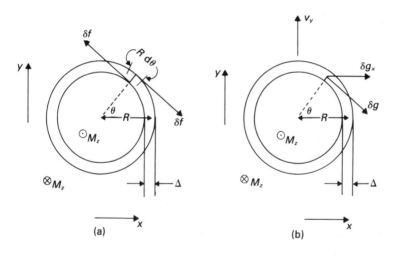

Figure 4.13 The bubble domain is moving along y and the angular momentum flux through the segment $R\,d\theta$ is considered in (a). The torque about this segment is equivalent to the force δg in (b)

It follows that there must be a change in the angular momentum density from $+M_s/|\gamma|$ to $-M_s/|\gamma|$, remembering that γ is negative, in a time $\Delta/v_y \sin\theta$, as the bubble domain moves forward, and that this change in angular momentum density takes place within the wall segment, volume $hR\,d\theta\,\Delta$. There must then be a torque about this volume element given by the rate of change of angular momentum, that is

$$\delta f \varDelta = [(2M_s/|\gamma|)/(\varDelta/v_y \sin \theta)] hR d\theta \varDelta \tag{4.64}$$

where the torque $\delta f \varDelta$ is illustrated in figure 4.13a and is shown with the correct sense for the signs of M_z and v_y, shown in figure 4.13 and with γ negative.

The small torque $\delta f \varDelta$, acting about a point on the wall is equivalent to a force, δg, acting at this same point, as shown in figure 4.13b, when δg satisfies the relationship

$$R \delta g = (R + \varDelta/2)\delta f - (R - \varDelta/2)\delta f \tag{4.65}$$

which follows from treating the bubble as a rigid body. If equation 4.64 is substituted into equation 4.65 and we take the integral of the x-component of δg, that is $\delta g \sin \theta$, around the entire bubble perimeter to find the total force in the x-direction the result

$$g_x = 2\pi M_s h v_y/|\gamma| \tag{4.66}$$

is obtained. The total force in the y-direction integrates to zero.

The force, g_x, is identical to the force which is produced on the bubble domain by the field gradient which we found in section 4.10.2 to be

$$\partial B_z/\partial x = -v_y/|\gamma|R^2 \tag{4.62}$$

This is immediately obvious when we recall the well known result that the force on a magnetic moment in a field gradient is simply the product of the moment and the gradient. The total magnetic moment of the bubble is its magnetic moment per unit volume, $2M_s$ relative to its surroundings, and its volume $\pi R^2 h$. The force is down the gradient and is thus

$$g_x = -(2M_s \pi R^2 h)(-v_y/|\gamma|R^2) \tag{4.67}$$

which is identical to equation 4.66. We shall refer to g_x as the gyrocoupling force in what follows, a term introduced by Thiele (1974).

4.10.4 Experimental Work on the Translation of S = +1 Bubbles

The experimental work on low S-number bubble translation will now be reviewed in order to see how well the theory of the two previous sections fits the experimental facts. We shall concentrate upon the deflection angles of bubbles which are thought to have the normal or $S = +1$ wall structure.

The tangent of the deflection angle, β shown in figure 4.12 is clearly given by the ratio $(\partial B_z/\partial x)/(\partial B_z/\partial y)$ which, from equations 4.62 and 4.63 is

$$\tan \beta = \frac{v_y}{|\gamma| R(v_y/\mu_w + 4B_c/\pi)} \tag{4.68}$$

This is not a very useful form because the magnitude of the velocity is more accurately measured and the magnitude of the field gradient, ∇B_z, is known accurately from the current flowing in the drive conductors. Coercivity and mobility are not known except as results from the same experiment which determines β. It follows that it is more accurate to neglect coercivity and use equation 4.42 to write

$$\mu_w = |v|/(R|\nabla B_z| \cos \beta) \tag{4.69}$$

where ∇B_z is the gradient shown in figure 4.12. Equation 4.69 will give an underestimate of the mobility, because we have neglected coercivity and thus attributed all of the gradient applied along the direction of motion, that is $|\nabla B_z| \cos \beta$, to overcoming the dissipation. It follows that if we set $B_c = 0$ in equation 4.68 and substitute equation 4.69 to obtain

$$\sin \beta = |v|/(|\gamma| R^2 |\nabla B_z|) \tag{4.70}$$

we have a reasonably accurate expression to use with the most directly available experimental data.

Equation 4.70 was given in the first paper to compare theory and experiment for an $S = +1$ bubble in translation (Slonczewski *et al.*, 1972) but was multiplied by a factor of 2. This factor of 2 was introduced as a footnote to earlier work (Slonczewski, 1972b) occurs in Thiele's work (Thiele, 1973b, 1974) and later work by Slonczewski (1974a, 1974b). The reason for its introduction is that these early papers treat the $S = +1$ bubble by means of a quite general theory which will be reviewed in section 4.10.6. This section continues with an account of the experimental work bearing this point in mind.

The first experimental papers which dealt with the translation of bubbles which were expected to have very low values of S, or to have $S = +1$, were concerned with garnet films supporting bubble domains $6 \mu m$ to $5 \mu m$ in diameter and with mobilities of the order of 10^5 m/s per T. When this data is available a simple estimate of the deflection angle may be made by neglecting coercivity in equation 4.68 and obtaining the very simple relationship

$$\tan \beta = \mu_w/|\gamma| R \tag{4.71}$$

which will give an optimistic value, an overestimate, for β.

For bubbles of 3.0μm to 2.5μm radius and mobilities of 10^5 m/s per T, equation 4.71 predicts a deflection angle of about 12° when we set $|\gamma| = 1.76 \times 10^{11}$ rad/s per T, the normal e/m value which is typical. It follows that only a very small deflection angle away from the direction of the gradient would be expected for $S = +1$ bubbles when the drive field was kept down to only 2 or 3 times the coercive limit, as discussed in section 4.10.1, in order to satisfy conditions of low drive.

This was certainly the case in the very detailed translation experiments published by Vella-Coleiro (1972). About fifteen different garnet compositions, all of high mobility, were studied using translation and no remarkable deflection angles were observed. This kind of result seemed to be confirmed when high speed microphotography was used in the same laboratory (Vella-Coleiro et al., 1975) on materials of similar composition: the translating bubble moved very close to the direction of the applied gradient.

Other workers in this field obtained rather different results. Slonczewski, et al. (1972), Voegeli and Calhoun (1973) and Malozemoff (1973b) all observed a deflection angle for a bubble domain in translation, which they identified as an $S = +1$ bubble, and the magnitude of this deflection angle agreed quite well with equations 4.70 or 4.71 when these expressions were multiplied by the factor of 2 mentioned above. Work by Smith et al. (1973), Bullock (1973) and de Leeuw and Robertson (1974) on 5 μm bubble domain films gave deflection angles more in agreement with equation 4.70 as it stands, however, and this was pointed out by O'Dell (1975). Bullock (1973) also found a very clear $1/R$ dependence for $\tan \beta$ from his data, as equation 4.71 would predict.

At this point we make contact with the remarks at the end of section 4.10.1 concerning the importance of data which comes from experiments using small bubbles. Equations 4.70 and 4.71 show that the deflection angle increases as the bubble size is decreased. At the time of these first bubble translation experiments, the development of the magnetic bubble domain device was concerned with using smaller and smaller bubble domains and the first report of a sideways deflection with micron-sized bubbles was given by Hu and Giess (1974) using four different LPE garnet films with Q-values between 4 and 2.1. In all cases the deflection angle of the bubbles was in very good agreement with equation 4.70 and this is particularly interesting because the LPE films were not ion-implanted and ion-implantation played an important role in a modified general theory (Slonczewski, 1974b) which might have explained these results at that time.

Bullock et al (1974) on the other hand, obtained deflection angles which were twice that predicted by equation 4.71 using submicron-sized bubble films of $(LaSm)_3Fe_5O_{12}$ and as this material has a very low mobility, because of the large amount of samarium it contains, the data should be well inside the low drive region. A very low mobility was also present in the amorphous metal films used by Potter et al. (1975) in their bubble translation experiments which were the first to show that the bubble domains in these interesting materials also showed the deflection effect. As Potter et al. (1975) indicated, however, the observed deflection angle of about 10° is much too big even when the factor of 2 is included in equations 4.70 and 4.71.

Further experimental work on garnet layers prepared for 5 μm bubble domain devices which found deflection angles about twice the value which would be predicted by equations 4.70 and 4.71 may be recorded as follows: Obokata *et al.* (1975) and Fontana and Bullock (1976) who studied the deflection angles in ion-implanted material over a wide range of temperature; Suzuki *et al.* (1975) and Suzuki and Sugita (1975) who looked at the deflection angles in permalloy coated material and made the very exciting discovery that bubbles with different deflection angles supported different kinds of surface domain in the permalloy layer; finally, Beaulieu and Calhoun (1976) who used what they called a deflecto-meter technique where the bubble domain was translated through a whole array of conductors, covering several bubble diameters, which was intended to make the measurement of the bubble deflection angle far more accurate—this study used ion-implanted material and included the effect of an in-plane field which caused elliptical distortion of the bubble domain.

Experimental work was also taking place at this time which did not put such emphasis upon the bubble deflection angle in translation and it is here that we begin to see the first suggestions that bubble translation might, in fact, be far more complex a phenomenon than had been previously thought. Josephs (1974) and Josephs and Stein (1974) questioned the very large statistical element that seemed to be present in bubble translation and which was clearly being filtered out in some way by some of the experimental methods being used. There was the very interesting observation by Malozemoff (1973b), Nakanishi and Uemura (1974) and Josephs and Stein (1974) of what came to be known as the turn-around effect (Maekawa and Dekker, 1976). This was that a bubble in translation which might have been classified as $S = +1$, and thus have no θ-dependent wall structure, as shown in figure 4.10, would be observed to re-verse its trajectory in translation, when the drive field was reversed, by following a completely new track for the first step or two. Such behaviour could only mean that the bubble concerned did have a θ-dependent wall structure after all and could not possibly be an $S = +1$ bubble. We shall consider this turn-around effect in section 4.12.

A true picture of what was really happening during the translation of low S-number bubbles began to emerge once high speed microphotography was applied to the problem. This is difficult experimentally because of the high velocities which are involved and the first pictures which were published (Vella-Coleiro *et al.*, 1975) caused great interest. Later work by Malozemoff and DeLuca (1975), Malozemoff *et al.* (1975) and Vella-Coleiro (1975, 1976b) showed that a bubble domain in translation would usually not stop moving when the gradient drive field was switched off but would continue to move forwards. Cases were observed in which the bubble continued to move for much the same distance as it had already travelled under the influence of the drive field. This effect would, of course, lead an experimentalist, using an ordinary microscope to observe only the initial and final positions of the translating bubble, to attribute too high a translational velocity to this bubble.

More details of the bubble translational motion came out as experimental techniques improved. Initial rapid wall motion had been observed by Zimmer *et al.* (1975) using high speed photography of simple bubble expansion and this gave direct confirmation of a phenomenon which had been inferred from translational experiments by Boxall (1974) and bubble collapse experiments by Vella-Coleiro (1974). Vella-Coleiro (1976a, 1976b) then applied high speed micro-photography to the bubble translation experiment and concentrated attention on what was happening at the beginning of the translational motion. It was observed that the bubble first began to move at a very high velocity and then, suddenly, a drop in the velocity would be observed. A clear connection between this kind of behaviour and the bubble overshoot phenomenon discussed in the previous paragraph was also found (Vella-Coleiro, 1976b). If there was no fast initial motion there would be no overshoot.

This kind of fast initial motion of domain walls is exactly what we would expect from section 3.7, where the observation of just the same phenomenon by de Leeuw (1974) in straight domain walls was discussed. Fast initial motion of bubble domain walls was confirmed in later experiments by Vella-Coleiro (1976c) using a very high time-resolution system to observe the radial contraction of an isolated bubble in contrast to the more difficult bubble translation experiment. Kleparskii *et al.* (1977) also took high speed photographs of bubbles in translation and observed fast initial motion with changes in deflection angle.

It is clear that fast initial motion followed by a drop in translational velocity or a change in the deflection angle must mean that the drive field being used is too large and the bubble wall structure is changing, discontinuously, as the bubble moves. When translation is actually being observed with a high speed microscope, it should be possible to reduce the drive and ensure that this kind of behaviour stops when the experiment is being conducted under the required low drive conditions.

The first attempt to do this appears to be by DeLuca *et al.* (1977a) who used the high speed microscope to observe the $S = +1$ bubble deflection angle and reduced the drive step by step until they could be confident that the symptoms of too high a drive field, these being fast initial motion and bubble overshoot, were negligible. The results then showed a bubble deflection angle which was in much better agreement with equations 4.70 or 4.71 as they are given here. The factor of 2, discussed at the beginning of this section, which would make equation 4.71 read

$$\tan \beta = 2\mu_w / |\gamma| R \qquad (4.72)$$

would have made agreement between experiment and theory very poor.

DeLuca *et al.* (1977a) took the expression for wall mobility from the simple

Landau–Lifshitz theory

$$\mu_w = |\gamma|\Delta/\alpha \qquad (4.73)$$

which was given in chapter 2 as equation 2.71, to write equation 4.72 as

$$\beta = \arctan(2\Delta/\alpha R) \qquad (4.74)$$

where we have corrected their typographical error. When the measured values of α from FMR were substituted into equation 4.74 a much larger value of β was given compared to the experimentally observed value. Again, this suggests that the factor of 2 in equation 4.74 may be incorrect and should be replaced by unity. The values of α for the garnets used by DeLuca *et al.* (1977a) are all just over 0.02 and this means that the experiments were well within the low drive field regime, which we are considering in this section, at the lowest drive fields applied.

To conclude this section on the experimental work, involving the deflection of $S = +1$ bubbles in translation, it is clear that there is a definite deflection effect which predominates over all other interfering effects and that this has the correct sign as far as the formal theory of section 4.10.2 is concerned and also from the simple physical point of view given in section 4.10.3. Quantitatively, things are not very satisfactory, however, because of the difficulty of keeping the experimental conditions within the low drive regime. The most recent work by DeLuca *et al.* (1977a), made particularly reliable by the use of high speed microphotography, supports the theoretical point of view adopted in sections 4.10.2 and 4.10.3.

4.10.5 Translation of Bubbles with Low S-numbers

We now turn away from the bubbles which have the simplest wall structure, the $S = +1$ bubbles shown in figure 4.10 in which M_θ simply circulates either clockwise or anticlockwise around the bubble wall, and consider the dynamics of the neighbouring states. These are the $S = 0$ state, the states with small negative S-numbers, on the left of figure 4.10, and the $S > 1$ states on the right. The static structure of these low S-number bubbles was discussed in section 4.9 and is shown generally in figure 4.7.

Hard bubble mobility was considered in sections 4.5 and 4.6 where it was shown that the essential feature of what might be called hard wall dynamics is that an applied field, which would normally be expected to drive the domain wall forward, is mainly effective in causing the pattern of M to propagate along the direction of the wall. In the case of a really hard bubble in translation, this effect resulted in the dynamic wall patterns shown in figure 4.4. The alternations of M in (r, θ) are no longer uniformly distributed around the bubble wall, as they are in the static case shown in figure 4.3, but become crowded on one side of the translating bubble domain.

A very similar effect must occur in a low S-number bubble but we now have to consider only a short segment of the wall. The effect of an externally applied magnetic field gradient is shown in figure 4.14 where the bubbles with $S = 0$ and $S = +2$ are considered. Only one alternation of M occurs around the wall in this case and when a gradient field is applied which would be expected to drive these bubbles up the diagram, the pattern of M within the wall begins to rotate. Because we have chosen, quite arbitrarily, to put the initial position of the two Néel lines in these bubbles on the trailing wall of the bubble, where the drive field B_z is negative, we have a clockwise rotation of the wall segment for the $S = 0$ case and an anticlockwise rotation for $S = +2$. This may be seen at once from the undamped equation of motion, equation 4.52, or from the argument which followed equation 4.28 when simple hard bubble wall mobility was considered.

As shown in figure 4.14, the initial response of the $S = 0$ bubble is that the two Néel lines move clockwise around the bubble wall. If the applied gradient is ideal, in that $B_z = 0$ across the horizontal diameter of the bubble shown in figure 4.14a, the Néel lines would be expected to come to rest on the left-hand side of the $S = 0$ bubble. This would be the case if there was no forward motion of the bubble itself. As it is, the Néel lines must come to rest at some stable riding point, shown as S in figure 4.14, where the drive field, B_z, is sufficient to keep the Néel lines moving at the translational speed of the bubble and give a state of dynamic equilibrium.

The final steady-state dynamic situation is shown in figure 4.15. The additional viscous drag associated with the Néel lines acts in such a way as to reduce the angle between the applied field gradient and the velocity vector in the case

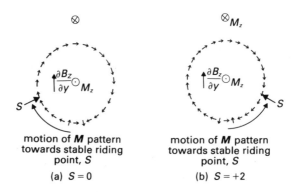

motion of M pattern
towards stable riding
point, S
(a) S = 0

motion of M pattern
towards stable riding
point, S
(b) S = +2

Figure 4.14 (a) A gradient, $\partial B_z/\partial y$, is applied to an $S = 0$ bubble and the Néel line cluster moves clockwise until it reaches a stable riding point. (b) The movement is anticlockwise in the $S = +2$ bubble under the same conditions

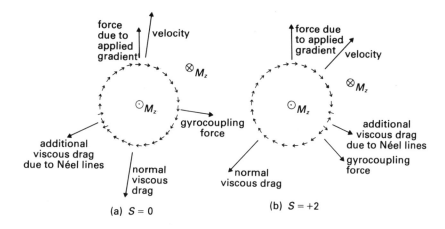

Figure 4.15 The balance between the viscous drag, gyrocoupling and applied forces in the moving bubbles: (a) $S = 0$, (b) $S = +2$

of the $S = 0$ bubble. For the $S = +2$ bubble, figure 4.15b, the effect of the additional viscous drag is to increase the bubble deflection angle.

This result applies when M_z is positive inside the bubble, as shown in figures 4.14 and 4.15, and this is the case where the $S = +1$ bubble has a positive deflection angle as discussed previously in connection with figure 4.12. For bubbles of the opposite polarity the argument is exactly the same but with all the signs reversed. It is important to note, however, that the steady-state translational motion of the bubble does not depend upon whether the bubble belongs to the χ^+ or χ^- states of figure 4.10 and also is independent of the initial position of the Néel lines within the wall. If this initial position had been chosen to lie upon the leading wall of the translating bubbles in figure 4.14, the end result would have been exactly the same because B_z is positive around the leading half of the bubble wall.

It is very clear, however, that the initial position of the Néel lines can have a dramatic effect upon the way in which the bubble may begin to move when in translation. In the same way, very complicated motion of the bubble may be seen when the direction of the applied gradient is reversed because the Néel line pairs will have to move right around the bubble wall to the opposite side before the steady-state conditions are satisfied again. This kind of behaviour was mentioned briefly in the previous section as the turn-around effect described and studied in detail by Maekawa and Dekker (1976) who pointed out that the only sure identification of the $S = +1$ state is that it not only shows the deflection angle defined by figure 4.12 but that it also shows no turn-around effect.

The deflection of the magnetic bubble which is produced by the viscous drag of the Néel line pair illustrated in figure 4.15, may be considered quantitatively using the low drive field model of this section which attributes a low field

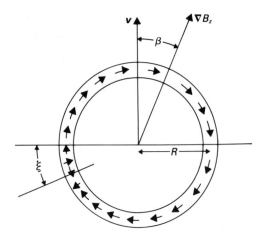

Figure 4.16 Defining the angles ξ and β

mobility $|\gamma|\Delta/\alpha$ to the normal bubble wall, equation 4.32, and a mobility $\alpha|\gamma|\Delta$ to a hard wall, equation 4.31.

Figure 4.16 shows the model we are using applied to the $S = 0$ state. The cluster of two Néel lines is shown at the stable riding point situated at angle ξ. The angle between the velocity vector and the applied field gradient is β. For the n used in section 4.4, the number of Néel lines is $2n$, that is, we are using n for the number of Néel line pairs. Using the results obtained in section 4.9 we may assume that the Néel lines in the cluster are separated by a distance $(\pi l/2Q^{1/2})$. The velocity of the cluster around the bubble perimeter will be $|v|\cos \xi$, when the bubble is in translation at velocity v, and it follows that the cluster must ride about a point where the local drive field, $|\nabla B_z|R \sin (\beta + \xi)$ is sufficient to cause a precession of the in-plane component of M in the cluster through an angle of π radians in $(\pi l/2Q^{1/2})/(|v|\cos \xi)$ s. This means that we must have

$$|\gamma|R|\nabla B_z|\sin (\beta + \xi) = (|v|\cos \xi)/(l/2Q^{1/2}) \qquad (4.75)$$

We now argue that the hard wall mobility, which is $\alpha|\gamma|\Delta$ from equation 4.31 when we neglect the term $(2n\Delta/D)^2$, implies a viscous drag on the hard wall segment, which is $(n\pi l/2Q^{1/2})$ long and has a radial component of velocity $|v|\sin \xi$, given by

$$f_d = (2n\pi Q^{1/2}hM_s/|\gamma|\alpha)|v|\sin \xi \qquad (4.76)$$

where we have used $l = 2Q\Delta$ from equation 4.8. This viscous drag will have a sideways component, $f_d \cos \xi$, figure 4.16, which is in balance with the sideways forces produced by the gradient, ∇B_z, and the gyrocoupling force, g_x from equation 4.66. It follows that

$$\sin \xi \cos \xi = (|\gamma|\alpha R^2|\nabla B_z|/nQ^{1/2}|v|)\sin \beta + (\alpha/nQ^{1/2}) \qquad (4.77)$$

There is also a component of viscous drag from the Néel line cluster which is in the direction of the velocity. This is $f_d \sin \xi$ and adds to the normal viscous drag of the entire bubble, $(2\pi RM_s h/\mu_w)|v|$, (O'Dell, 1974), which, using $\mu_w = |\gamma|\Delta/\alpha$, yields the relationship

$$\sin^2 \xi = (|\gamma|\alpha R^2|\nabla B_z|/nQ^{1/2}|v|)\cos \beta - (2\alpha^2 RQ^{1/2}/nl) \qquad (4.78)$$

If equations 4.77 and 4.78 are substituted into equation 4.75, the angle β and the velocity $|v|$ may both be eliminated to leave an expression

$$\tan \xi = (1 + l/2RQ^{1/2})/(\alpha Q^{1/2} + nl/2\alpha R) \qquad (4.79)$$

We may assume that $\alpha \ll 1$ and that $l \ll 2RQ^{1/2}$. It then follows that equation 4.79 may be simplified to

$$\tan \xi = 2\alpha R/nl \qquad (4.80)$$

showing, with reference to figure 4.16, that the Néel line cluster rides very close to the flank of the bubble, ξ being very small even when n takes its smallest value, unity. The model used by Thiele (1974), which is quite different to the one being used here, also showed that the angle ξ was very small.

The small value of ξ allows a remarkable simplification in the remainder of this calculation because it allows us to write $\tan \xi \approx \sin \xi \approx \xi$ and $\cos \xi \approx 1$ and then the sideways force due to the viscous drag of the Néel line cluster is simply

$$f_d \cos \xi = 4\pi hQ^{1/2}M_s R|v|/|\gamma|l \qquad (4.81)$$

Equation 4.81 shows that this force is independent of n and is very much larger than the gyrocoupling force

$$g_x = 2\pi hM_s|v|/|\gamma| \qquad (4.82)$$

given by equations 4.66 and 4.67. It follows that the last term in equation 4.77 which came from the gyrocoupling force, may be neglected and equation 4.78 divided by equation 4.77 to give

$$\cot \beta = (\alpha Q^{1/2} + 2\alpha R/nl) \qquad (4.83)$$

as an expression for the bubble deflection angle, β. For normal values of Q and small values of n, the second term in equation 4.73 will dominate and we may write

$$\tan \beta = nl/2\alpha R \tag{4.84}$$

which has the interesting form

$$\tan \beta = (\mu_w/|\gamma|R)(nQ) \tag{4.85}$$

when we replace α by $|\gamma|\Delta/\mu_w$, equation 4.75, and use $l = 2Q\Delta$, equation 4.8.

It will be recalled that the bubble deflection angle which is expected from the gyrocoupling force alone is $\tan^{-1} (\mu_w/|\gamma|R)$, equation 4.71, and that the experimentally determined angles were often much larger. Equation 4.85 now gives a possible origin for this kind of experimental result which might be added to the remarks given in section 4.10.4 in cases where there is some reason to suspect, because of the presence of turn-around effect for example, that the bubble wall may contain a small number of Néel line pairs and not be a pure $S = +1$ state.

Finally, we may use equation 4.84 to obtain an expression for $\cos \beta$ and then use this in equation 4.78, with equation 4.80 in the small angle approximation, to get an expression for $|v|$. When we use $\mu_w = |\gamma|\Delta/\alpha$, equation 4.73, and $l = 2Q\Delta$, equation 4.8, this expression may be written

$$|v| = \frac{\mu_w |\nabla B_z|R}{(1 + 2R/nlQ^{1/2})(1 + n^2 l^2/4\alpha^2 R^2)^{1/2}} \tag{4.86}$$

Equation 4.86 shows how the bubble velocity falls as the number of Néel line pairs, n, is increased, relative to the velocity deduced from the simple constant mobility model, which is $\mu_w |\nabla B_z|R$ when coercivity is neglected. The reduction in bubble velocity predicted by equation 4.86 is of the same order of magnitude as that observed experimentally (Voegeli and Calhoun, 1973) when values of $\alpha \approx 10^{-2}$ are used with the other material parameters given. The model is too crude, however, for any good numerical agreement to be expected in the case of velocity.

The calculations which have been given above have added a certain amount of numerical detail to the qualitative picture presented at first by figures 4.14 and 4.15. The order of the deflection angle and the velocities which the model predicts are certainly the same as those observed experimentally with bubbles which show a turn-around effect suggesting that they are not pure $S = +1$ states but are bubbles which do contain a few Néel line pairs within their walls.

4.10.6 Theoretical Problems

The development of a general theory of magnetic bubble domain dynamics was described in section 4.2 and an account of this may be found in the papers by

Thiele (1974) and Slonczewski (1979). This general theory would double the gyrocoupling force, given here by equation 4.66, but a much more important point is that the general theory attributes a gyrocoupling force, not only to the whole bubble domain but also to the Néel line cluster in the wall. Because this result follows from considering only the topology of the bubble wall, the entire wall of an $S = +1$ bubble domain is of exactly the same weight in the argument as one pair of Néel lines. It follows that the gyrocoupling force due to one pair of Néel lines has the same magnitude as the gyrocoupling force on the entire $S = +1$ bubble wall.

This results in the $S = 0$ bubble domain, shown in figure 4.10, having zero deflection angle in the bubble translation experiment, the gyrocoupling force from the two Néel lines balancing the gyrocoupling force from the domain itself. The generalisation of equation 4.70 given by Slonczewski *et al.* (1972) is, in the symbolism used here

$$\sin \beta = 2S|v|/(|\gamma|R^2|\nabla B_z|) \tag{4.87}$$

and this predicts that the value of β for the $S = +1$ bubble must be larger than that for the $S = -1$ bubble because of the lower value of $|v|$ in the $S = -1$ case. This behaviour was observed by Voegeli and Calhoun (1973) whose experimental results are in quite good agreement with equation 4.87, particularly for the states $S < 0$. Beaulieu and Calhoun (1976), on the other hand identified $S = +1$ and $S = -1$ bubbles, in a film of almost the same composition as the one used by Voegeli and Calhoun (1973), and measured deflection angles which were very similar for these two states: 32° and 28°.

No deeper consideration of the general theory will be given here because it seems certain that there must be at least three developments before this can be useful. The first development which is needed is to explain the physical origin of the gyrocoupling force on a Néel line pair, as we have attempted to do here for the gyrocoupling force on the entire bubble in section 4.10.4. Secondly, the role of the dissipation caused by the moving Néel lines appears to be lost in the work of Thiele (1974) and Slonczewski (1974a), Slonczewski and Malozemoff (1978). This must play a very important part in causing bubble deflection, as illustrated in figure 4.15. Finally, a critical look at the experimental facts using high-speed microphotography is essential and the work of DeLuca *et al.* (1977a) marks the beginning of this approach.

Another important theoretical problem is the part played by coercivity in the simple translation of magnetic bubbles. This was discussed at length in section 4.10.1 from the experimentalist's point of view: the finite coercivity of our materials makes it very difficult to conduct experiments within the low drive field regime. How to include coercivity in the dynamic theory, however, is a question which is still open, although much valuable discussion of this problem is given by Thiele (1974) and Slonczewski (1974a), and in their earlier papers, while coercivity is brought into later work by Malozemoff (1976) which is more concerned with the high drive regime, considered here in section 4.13.

Finally, the problem of coercivity and very low drive should be mentioned. This is a well established problem in magnetism, often referred to as 'creep' and an excellent review of this general problem with some new theoretical work has been given by Baldwin *et al.* (1977). The problem is that a drive field below what would normally be considered the coercive threshold does, in fact, cause very slow wall motion. Barbara *et al.* (1977) have looked at this problem, experimentally, using bubble translation.

4.11 Bubble Dynamics in Ion-implanted and Multilayer Garnet Films

The suppression of hard magnetic bubbles by means of ion-implantation, and by other techniques, was discussed at length in section 4.8. The generation of bubbles with large magnitude S-numbers is suppressed by means of these techniques and it follows that we would expect all the bubbles in an ion-implanted film to behave in the manner of $S = +1$ bubbles and show a deflection angle in translation of the kind shown in figure 4.12.

Experimental work by Bullock (1973), however, had shown that a number of different bubble states could, in fact, co-exist in an ion-implanted film provided an in-plane field was applied. This fact soon led to the suggestion that information could be stored by using bubble domains belonging to different states (Voegeli *et al.*, 1974). Such a data store could have very high density because the bubbles might be formed into a close-packed hexagonal lattice. There was also the possibility of using more than just two states for data storage, for example, the $S = +1$, 0 and -1 states might be chosen to represent numbers to base 3. All three states have mobilities of the same order of magnitude, as shown by figure 1.12, and may be clearly distinguished by means of their deflection angles.

This idea became a working device, (Hu *et al.*, 1978), after Hsu (1974) discovered a method of writing the $S = 0$ and $S = +1$ states in ion-implanted films. Hsu (1974) found that both states were stable when an in-plane field was applied, which had to lie within a critical range. The $S = 0$ state would always be found after a pulsed magnetic field was applied to a bubble in the plane of the film, while the $S = +1$ state would follow the application of a pulsed field added to or subtracted from the bias field supporting the bubble in question. These pulses were quite short, ≈ 100 ns, and of the order of $\mu_0 M_s$.

The ion-implanted layer, lying on top of the film used in such an experiment, has its magnetisation lying in-plane and Hsu (1974) explained his results by considering the effect of the magnetic domains which could form in this ion-implanted layer. These planar domains have received considerable experimental attention (Puchalska *et al.*, 1977) since they were first discussed by Wolfe and North (1974). Because the ion-implanted layer is on top of the epitaxial garnet film it may be easily accessed and the planar domains within it decorated with magnetic colloid. Such decoration shows that it is possible to have a domain in the ion-implanted layer which practically covers the area occupied by the bubble and, because M within the ion-implanted layer lies in-plane, this domain tends to support a bubble domain in the main film having the very special kind of wall structure shown in figure 4.17. This wall is formally identical to the $S = 0$ structure, shown in figure 4.10, when only the topology or winding number, S, as defined in section 4.4 is considered. Because the general theory of bubble dynamics only takes account of the value of S, and is not concerned with the details of the wall magnetisation, the hypothesis that the structure shown in figure 4.17 is the one obtained by Hsu (1974) as a result of an in-plane pulsed field is supported by the fact that the bubble obtained by in-plane field pulsing does, subsequently, translate at a negligible angle to the applied gradient. From the point of view taken here, however, it would be suggested that the problem is more complex and would involve the translational motion of the planar domain in the ion-implanted layer as well. The effect of this on the bubble motion could easily swamp out the small deflection angle predicted here by equation 4.70 for a bubble wall, such as the one illustrated in figure 4.17 which contains no Néel lines in a cluster.

In what follows, we shall label the two bubble states defined by the writing technique of Hsu (1974) as $S = 0$ and $S = +1$, although it is clear that this may be an oversimplification. Work on the properties of these two bubble states in ion-implanted films under constant in-plane field gave further interesting results. Hsu *et al.* (1975) looked at the speed difference between the two states and found little difference at low drives but at higher drives the $S = 0$ state was found to be more mobile. Temperature stability of the bubble states and transitions between bubble states were studied by Obokata *et al.* (1975). Beaulieu *et al.* (1976) gave a very detailed account of the conditions under which transitions

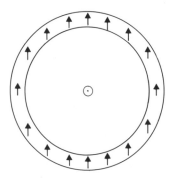

Figure 4.17 The bubble domain wall structure which may be found in ion-implanted films, films having an in-plane anisotropy or when an in-plane field is applied

from one state to the other could be observed to occur and a model to predict the critical in-plane fields at which these transitions happened was given by Okabe (1978) and was in good agreement with the experimental facts.

All this work on transitions between states, or state switching, involved the new concept of the Bloch point which had been brought into bubble dynamics by Slonczewski (1974b) to explain the small deflection angles which were observed for bubbles translating in certain samples of material. The term 'Bloch point' was attributed by Slonczewski (1974b) to Feldtkeller (1965), who does not in fact use this term but discusses a number of possibilities for a micromagnetic singular point. In bubble dynamics, the idea is to introduce a more complicated structure for the transition between parts of the bubble domain wall which have different screw-senses. This idea is shown in figure 4.18 where we have adapted the diagram given by Feldtkeller (1965) for the singular point which would occur in a similar transition in the wall between two domains having in-plane magnetisation; for example, in a thin permalloy film. The situation is quite different in a bubble domain film because of the magnetic field which circulates around the outside of the wall region in (r, z), discussed at length in chapter 3, figure 3.2. In figure 4.18, (y, z) is equivalent to (r, z).

Figure 4.18 shows a domain wall separating two domains, of the kind which would exist in a bubble domain film, and the screw-sense of the wall is clockwise on the right-hand side of the diagram but anticlockwise on the left. Where these two different screw-sense walls join, we no longer have a simple Néel line, as we did in the previous sections of this chapter, but have a Néel line in the top half of the film which has M directed along $+y$ while the bottom half of the film has a Néel line with M directed along $-y$. In this way, M has been made to lie parallel to the B field which circulates outside the wall in (y, z), the field of figure 3.2.

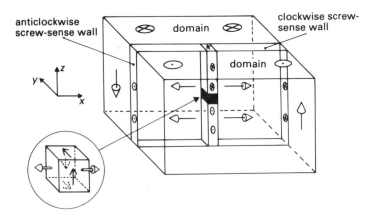

Figure 4.18 The diagram given by Feldtkeller (1965) showing a micromagnetic singular point in a domain wall has been adapted here for the bubble wall case

The magnetostatic energy of this double Néel line is thus lower than the magnetostatic energy of the simple Néel line but the exchange energy is higher. If the film is thick enough, the double Néel line will have a lower total energy than the simple Néel line (Hubert, 1976). The complicated spin field around the centre of figure 4.18 is shown in more detail in the bottom left-hand corner of this diagram. It should be noted that this is not a true micromagnetic singularity, in the sense Feldtkeller (1965) described, because it has an axis of rotational symmetry. This has been discussed by Slonczewski and Malozemoff (1978).

The kind of wall structure shown in figure 4.18 is a perfectly satisfactory hypothesis as an isolated entity but things are not so clear when we consider it as part of a closed bubble domain wall. There must obviously be a second transition in the main wall screw-sense before the bubble wall may be closed upon itself and we would expect the two transitions to cluster because this would be the lowest magnetostatic energy configuration—a point considered in detail in section 4.9. However, the two transitions may remain separated and diametrically opposed to one another if an in-plane field is applied to the bubble domain film. This proposal is shown in figure 4.19 and has the additional interesting feature that the double Néel line, originally shown in figure 4.18 and now situated at the top of figure 4.19, need no longer be symmetric about the central plane of the film, $z = 0$, so that it has zero net magnetisation along y, but may become magnetised in the direction of the in-plane field, B_y (Slonczewski, 1974b, Hasegawa, 1974). Similarly, the in-plane field supports the polarity and position of the simple Néel line, shown at the bottom of figure 4.19, and also supports the direction of M within the walls on either side.

From the point of view of the general theory of bubble dynamics (Thiele, 1974, Slonczewski, 1974a), which considers only the topology of the wall, the bubble domain illustrated in figure 4.19 belongs to the state $0 < S < 1/2$, depending upon the magnitude of B_y. For $B_y \approx 0$ we have $S \approx 1/2$. Because the

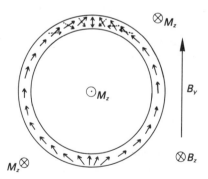

Figure 4.19 An in-plane field is shown supporting a bubble having an isolated single Néel line diametrically opposing an isolated double Néel line

dynamic behaviour is determined by the value of S, in the general theory, this argument was used to explain the small deflection angles which had been observed for bubbles expected to be $S = +1$, discussed in section 4.10.4, and Hasegawa (1974) gave more experimental data from this new point of view.

The dynamic behaviour of a bubble domain which does have the hypothetical wall structure shown in figure 4.19 may be discussed from the point of view of section 4.10.5. If we imagine a gradient field applied to the domain of figure 4.19, of such a polarity as to drive it in the same direction as B_y, which we assume to be well below $\mu_0 M_s$, then it is clear that the simple Néel line shown at the bottom of the diagram will begin to move clockwise around the bubble domain wall. This follows from the argument which was applied to figure 4.14a. The final stable riding point for this Néel line, however, is now influenced by the in-plane field because the magnetostatic energy of the entire wall depends upon the Néel line position. As the lower Néel line in figure 4.19 moves clockwise, some of the left-hand side of the wall must become magnetised in the opposite direction to B_y. This is a higher energy state and it follows that the Néel line will move a smaller distance around the circumference the larger B_y is made.

If we now consider the double Néel line at the top of figure 4.19, quite a different picture emerges. The part of this structure, the top part, which lies parallel to B_y will try to move anticlockwise while the bottom part will try to move clockwise. This follows because the local B_z from the gradient drive will be positive around the leading half of the bubble wall while it was negative around the trailing half which we considered in the previous paragraph. If the in-plane field is weak, so that the two halves of the double Néel line are nearly balanced, no net motion will result but the double Néel line will be distorted and store energy. This distorted structure now has to move forward with the moving bubble domain.

If the in-plane field in figure 4.19 is high enough to make the double Néel line become magnetised along the direction of B_y, the double line will move anticlockwise and add to the deflection produced by the simple Néel line in the trailing wall. It is now obvious why the deflection angle should be influenced by the magnitude of the in-plane field and an interesting account of some experiments dealing with the dependence of the bubble translation deflection angle, in an ion-implanted film, upon the intensity of the applied in-plane field has been given by Josephs *et al.* (1975). The results were explained by proposing that the bubble in question contained two of the double Néel lines of the kind shown in figure 4.18. Further work along these lines was described by Beaulieu *et al.* (1976), who developed a new notation for describing the possible wall states which bubble domains could have in ion-implanted films. This notation will be introduced in the next section. Beaulieu *et al.* (1976) confirmed that it was possible to observe bubbles changing state discontinuously, an effect which was discussed at the beginning of this section for the simplest case $S = +1 \rightleftharpoons S = 0$. It became clear, however, that switching was taking place between far more complicated states and this idea is leading us towards the last section of this chapter where we shall consider bubble dynamics under high drive field conditions when the bubble may change its state during motion and even take up a state which is only stable dynamically. Before considering this, however, we must look at a further kind of low drive field dynamics which can give a considerable amount of new data about the existence of more complex wall states.

4.12 Automotion

It should be emphasised again that this chapter on bubble domain dynamics has, up to now, been concerned with what we have been calling the low drive field regime. This is the regime within which we may assume that the initial state of the bubble, that is its static structure, is a fairly accurate model for its dynamic structure and that, as a consequence of this, its final state, after motion, will be identical to its initial state.

In the next and last sections of this chapter we shall look at what happens when the low drive field assumption is clearly no longer tenable and dramatic changes in the bubble state may occur during motion, either reversibly or irreversibly. Before doing this, however, we shall look at a very different kind of bubble motion which belongs to the low drive field regime but does depend upon the existence of quite complicated bubble wall states. This kind of motion has been termed 'automotion' by Argyle *et al.* (1976a) because the bubbles appear to move on their own account, there being no obvious gradient drive field applied.

The experiments described by Argyle *et al.* (1976a), more detail being given in a later paper (Argyle *et al.* 1976c), involved the standard bubble translation experiment, figure 1.11, in which a bias field gradient could be applied to a bubble by passing currents through two parallel conductors laid down upon the bubble film. To this set-up, an in-plane field was added, parallel to the direction of the gradient drive field conductors, and provision was also made for adding a pulsed or sinusoidally varying bias field to the normal constant bias field supporting the bubbles under investigation.

Using this apparatus with a bubble domain film of the $(GdYTm)_3(FeGa)_5O_{12}$ composition, Argyle *et al.* (1976a) found that bubble domains which could be assigned to the $S = +1$ state, using the translation experiment to observe their deflection angle, could also be made to move if a small in-plane field of 1.2 mT was applied together with an expanding bias field pulse of 10 μs duration and 1 mT to 2 mT in amplitude. Not all the $S = +1$ bubbles showed this 'automotion' and those which did showed a whole range of directions of motion relative to the direction of the in-plane field. Similar results were obtained in films of other compositions (Argyle *et al.*, 1976a) and in ion-implanted films (Argyle *et al.* 1976c).

We first consider bubbles which move perpendicular to the applied in-plane field when the bias field is pulsed or modulated. Clearly, this kind of auto-motion must be due to some non-linear dissipative term in the equations of motion because no change in the bubble energy takes place, when it moves in this way, when we compare initial and final positions. The non-linear dissipative term is almost certain to be coercivity and this, coupled with the unusual $S = +1$ bubble states shown in figure 4.20, was used by Argyle *et al.* (1976a) to explain their experimental results.

Two possible bubble states are shown in figure 4.20 which, like the bubble of figure 4.19, both have walls of opposite screw-sense upon either side, these walls being supported by the in-plane field, B_y. The two transition regions, or Néel lines in these bubbles, are diametrically opposed and have opposite winding sense so that, if B_y is removed, the two Néel lines would annihilate one another if they were to cluster in the usual way.

By following the argument given in section 4.5, immediately after equation 4.28, it is clear that an expanding bias field pulse applied to the bubbles shown in figure 4.20 will cause all the M which lie in (x, y) to begin to rotate anti-clockwise. This will mean that both Néel lines shown in figure 4.20a will move around the circumference of the bubble towards the right-hand side. The two Néel lines shown in figure 4.20b, on the other hand, will move to the left. This, of course, only occurs while the bubble is expanding towards its new equilibrium diameter. When equilibrium is achieved, the magnetostatic field of the bubble is again equal and opposite to the sum of the effective wall field and the bias field so that the two Néel lines will return to the diametrically opposed positions shown in figure 4.20.

To understand the origin of the automotion, consider figure 4.20a. As this bubble expands and the Néel lines move to the right-hand side, producing a lower mobility for the right-hand side, the bubble must expand with the left-hand wall moving more rapidly. The bubble centre thus moves to the left and, because coercivity always opposes the wall motion, just like Coulomb friction,

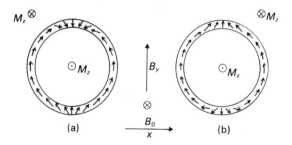

(a)   (b)

Figure 4.20 An in-plane field is shown supporting two bubble domains which have diametrically opposed Néel lines, these lines having opposite winding sense

this motion will not be cancelled out when the Néel lines relax back to occupy diametrically opposed positions in the larger bubble. The drive pulse now terminates, which is equivalent to a negative pulse in B_z and will cause the Néel lines in the bubble wall of figure 4.20a to move to the left-hand side of the contracting bubble. It is now the right-hand wall which will move more rapidly during the contraction so that, again, the bubble centre moves to the left. Clearly, this kind of motion is independent of the sign of the drive pulse, and this is confirmed experimentally (Argyle *et al.* 1976a).

A similar argument applied to the bubble shown in figure 4.20b shows that bias field modulation will cause this bubble to move to the right. Argyle *et al.* (1976a) named the bubble states shown in figures 4.20a and b the σ_- and σ_+ states, respectively, the sign being chosen according to whether (v_x, B_y, B_0) formed a left-handed or a right-handed set of coordinates. Note that we have maintained our convention of having M positive inside the bubble in figure 4.20. Argyle *et al.* (1976a) chose to have M_z negative inside the bubble but this is identical to a two-fold rotation about y and does not change any definitions or results.

An alternative notation, introduced by Beaulieu *et al.* (1976) labels the more complex bubble states (S, L, P) where S is the state-number, defined in section 4.4, L is the number of Néel lines, double or single, and P is the number of 'Bloch points', discussed here in connection with figure 4.18. This notation is not complete, however, because both states shown in figure 4.20 would be $(1, 2, 0)$ and must be distinguished as $(1, 2, 0)_-$ and $(1, 2, 0)_+$.

Before turning to the other kinds of automotion described by Argyle *et al.* (1976a, b, c) we should look at some very interesting and elegant experimental work which supports the hypothetical bubble wall states illustrated in figure 4.20 very strongly. This is the work of Dekker and Slonczewski (1976) in which the transitions between the $(1, 2, 0)_+$, $(1, 2, 0)_-$ and the two $S = +1$ states of figure 4.10 were studied.

To illustrate the kind of experimental work reported by Dekker and Slonczewski (1976), consider the $S = +1$ bubble shown in figure 4.21a. This is labelled $(1, 0, 0)_-$ because the circulation of M_θ around the wall, using the right-hand screw rule,

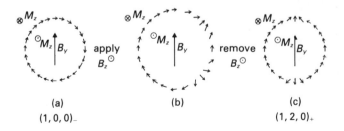

(a)   (b)   (c)

$(1, 0, 0)_-$ $(1, 2, 0)_+$

Figure 4.21 One of the state switching processes studied by Dekker and Slonczewski (1976)

is opposed to the direction of M_z inside the bubble. We now apply a small in-plane field, B_y, perhaps $B_y \approx \mu_0 M_s/10$. According to our work in chapter 3, such a small in-plane field will only affect those parts of the bubble wall where B_y is mainly parallel to the plane of the wall and its effect will be to increase the critical drive field for maximum wall velocity where B_y is parallel to M_θ. If we now apply a bias field pulse of such a polarity as to expand the unichiral bubble shown in figure 4.21a it is clear that the right-hand side of the wall will break-down first and, if the correct intensity and duration of bias field pulse is chosen, we shall effect the state transition $(1, 0, 0)_- \rightarrow (1, 2, 0)_+$ shown in figure 4.21.

If a contracting, or negative, bias field pulse had been applied, in the experiment described above, the same side of the bubble shown in figure 4.21a would have switched its screw-sense but the two Néel lines generated would have had M pointing radially inwards. That is, we would have switched the $(1, 0, 0)_-$ state to $(1, 2, 0)_-$, the state shown previously in figure 4.20a.

Now the unichiral bubbles $(1, 0, 0)_+$ and $(1, 0, 0)_-$ may be identified because they show no automotion and translate at the small deflection angle, defined in figure 4.12 and by equation 4.70, with no turn-around effect. However, we must be able to distinguish between these two unichiral states before the hypothetical switching between bubble states can really be tested experimentally. Further consideration of the hypothetical wall structures shown in figure 4.20 suggests a way out of this problem. This is to ask what happens if we remove the in-plane field shown supporting the $(1, 2, 0)_-$ and $(1, 2, 0)_+$ states in figure 4.20, and then apply a pulsed bias field. As in automotion, this will cause the two Néel lines to move towards one another and, with no in-plane field, there is no longer any reason why the Néel lines should not run into one another and annihilate to produce a unichiral bubble.

Following the argument of section 4.5, equation 4.28, again, an expanding bias field pulse will cause the Néel lines within the bubble wall shown in figure 4.20a to move to the right and, if they then collide and annihilate, a $(1, 0, 0)_-$ state, figure 4.21a will result. The opposite chirality, $(1, 0, 0)_+$, would be obtained if a contracting bias field pulse were used. We now have a complete and closed set of experimental tests where we may switch between two identifiable bubble states which show automotion via one state which, while it is not intrinsically identifiable, may be identified uniquely from the state it is created from and the method it is created by, coupled with the properties of the state which it is switched to and the methods which are used to implement that switch. This is summarised in table 4.1 where some of the experimental tests of Dekker and Slonczewski (1976) are listed.

In their experiments, Dekker and Slonczewski (1976) brought about the reversal of the screw-sense of the walls they were working with by using variable rise time pulses. This is a far easier method to control compared to adjusting the amplitude and duration of the drive pulse. Because the effective drive field at the wall of an expanding or contracting bubble domain begins to fall immediately the wall begins to move, it is possible to choose the rise time of the drive pulse in

Table 4.1 Eight state switching sequences verified by Dekker and
Slonczewski (1976)

Establish Initial State	B_y Off Pulse B_z	Check Unichiral State	B_y On Pulse B_z	Check Final State
$(1,2,0)_+$	$+$	$(1,0,0)_+$	$+$	$(1,2,0)_-$
$(1,2,0)_-$	$+$	$(1,0,0)_-$	$+$	$(1,2,0)_+$
$(1,2,0)_+$	$+$	$(1,0,0)_+$	$-$	$(1,2,0)_+$
$(1,2,0)_-$	$+$	$(1,0,0)_-$	$-$	$(1,2,0)_-$
$(1,2,0)_+$	$-$	$(1,0,0)_-$	$+$	$(1,2,0)_+$
$(1,2,0)_-$	$-$	$(1,0,0)_+$	$+$	$(1,2,0)_-$
$(1,2,0)_+$	$-$	$(1,0,0)_-$	$-$	$(1,2,0)_-$
$(1,2,0)_-$	$-$	$(1,0,0)_+$	$-$	$(1,2,0)_+$

such a way as to keep the effective drive field almost constant. The actual value of the effective drive field is then set by varying the rise time and can be set very accurately in this way.

This experimental refinement enabled Dekker and Slonczewski (1976) to produce what they termed 'chiral reversal' by applying an expanding bias field pulse, with the correct rise time and final amplitude, to the $(1, 0, 0)_+$ or $(1, 0, 0)_-$ states, with no in-plate field applied. It was then possible to perform eight more experiments in which the chirality of the bubble shown in the centre column of table 4.1 was switched and the experiment continued as before. It was confirmed that this procedure switched the sign of the final $(1, 2, 0)$ state, as inspection of table 4.1 would lead us to expect.

To return to Argyle *et al.* (1976a, 1976c), the $(1, 2, 0)_+$ states were not the only bubbles observed to show automotion. States were found which showed automotion along the direction of the applied in-plane field, either parallel to it or antiparallel, and also at angles of 45°, 135°, 225° and 315°. These results were explained by proposing that the Néel lines within the bubble domain walls contained the kind of transition shown in figure 4.18, the 'Bloch point', but more experimental work is needed before this can be accepted. The $(1, 2, 0)_+$ states themselves are subject to certain interpretational difficulties in that very similar automotion properties should be found in bubbles having clusters of an odd number of Néel lines in place of the single ones shown in figure 4.20. This was first suggested by Argyle *et al.* (1976a) and some experimental evidence which seems to support the possibility has been given by Iwata *et al.* (1979) who observed states in automotion perpendicular to the direction of the in-plane field, but at two different speeds, and suggested that the (0, 4, 0) state was responsible because the faster bubble in automotion showed no deflection angle in translation.

The work involving automotion is of great importance because it marks the beginning of a systematic attempt to confirm or deny the hypothetical wall structures which have been proposed in order to explain many phenomena in bubble domain dynamics. Other work involving automotion is the work on bubble lattices by Argyle *et al.* (1976b) and work on the turn-around effect by Maekawa and Dekker (1976) which was discussed in section 4.10.4. Recently Ju and Humphrey (1979) have combined automotion and high-speed micro-photography, using a technique to determine the bubble state which has been described by Gallagher *et al.* (1979) involving an observation of the true velocity of the bubble wall at various points around its circumference. This work has raised a number of unanswered questions and future developments in this experimental field promise to be of great interest.

4.13 Bubble Translation under High Drive Fields

This chapter closes with a discussion of bubble dynamics under drive fields which are much larger than the drive fields considered in previous sections. This is a continuation of the work covered in chapter 3, section 3.9, where the experimental facts given in figure 3.9 showed that a domain wall could move forward at a very high velocity but that this high velocity mode of motion would collapse, once the drive field was increased above a certain threshold, and the wall would then move forward with a much reduced mean velocity that was termed the saturation velocity.

When a large in-plane field is applied to the bubble domain film, as it was in all the work of chapter 3, this kind of behaviour is fairly straightforward. Without the in-plane field, matters become more complicated and more detail must now be added to the simple picture presented in chapter 3 which led to our result, equation 3.27, for the saturation velocity of the domain wall under conditions of zero in-plane field.

Figure 4.22a shows the initial or static structure expected in a straight domain wall in a bubble film; a combination and simplification of figures 3.1 and 3.3. About the central plane of the film the wall may be assumed to have the simple Landau–Lifshitz structure but near the top and bottom surfaces of the film M turns so that it lies parallel to the magnetostatic field coming from the domains upon either side of the wall. This is the field that was shown in figure 3.2.

Figure 4.22b shows what must happen to this structure when a drive field, B_z, is applied to make the wall move to the right. According to the work of chapter 3, the magnetostatic field at the bottom surface is of the correct sign to support high velocity motion when the wall has a clockwise screw-sense and this was chosen, quite arbitrarily for our diagram, in the first place. It follows that the wall can move forwards at the bottom of the film with the twist in the

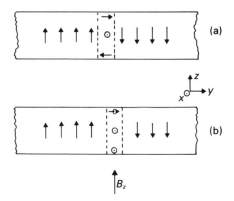

Figure 4.22 The initial, static, wall structure in a bubble domain film (a) must unwind when a drive field, B_z, is applied as in (b). The wall width is exaggerated in this diagram, and in those which follow, for clarity

wall structure unwinding, as shown in figure 4.22b. At the upper surface of the film the magnetostatic field is positive and it was shown in chapter 3, section 3.8, that this sign of in-plane field would stabilise a domain wall with an anticlockwise screw-sense. It follows that the top part of the wall will try to take up this anti-clockwise screw-sense but, as shown in figure 4.22b, this means that the twist in the wall structure through the upper thickness of the film may have to increase and the resulting exchange field will have a very similar stabilising effect to the in-plane magnetostatic field which is already present at the top surface of the film.

It may be concluded from this argument that the velocity of a straight domain wall in a bubble film should be able to increase to a value very close to the Walker velocity, found in chapter 2, section 2.11, to be

$$v_W = |\gamma| \mu_0 M_s \Delta / 2 \qquad (4.88)$$

before the dynamic structure shown in figure 4.22b collapses and the saturation velocity, discussed in chapter 3, section 3.9, takes over. This conclusion follows from the results which led to figures 3.7 and 3.8. These show that the peak velocity of the wall should be close to v_W when the in-plane field is as low as $\mu_0 M_s$, which is the value of the magnetostatic field along y at the surface of the film to a close approximation (O'Dell, 1978a). It is only possible to estimate the velocity because the theory given in chapter 3 was specifically restricted to in-plane fields well above $\mu_0 M_s$.

Attempts to treat the dynamics of straight domain walls in bubble domain films under conditions of weak in-plane field have been made by Gurevich (1977) and by Nedlin and Shapiro (1977) but the assumptions made in these theoretical treatments make it difficult to apply them to a practical case and obtain a more accurate picture than the one given here. The conclusion that the wall should move at a velocity close to v_W when a fairly large drive field, that is one well above $\alpha\mu_0 M_s/2$, is applied and that the velocity should then drop or become fluctuating is supported by some experimental evidence. Vella-Coleiro (1976c) investigated simple bubble contraction, which should be very similar to simple straight wall motion, using a high speed microscope with a space–time resolution of $0.1\,\mu$m–1 ns. The garnet layer used had the parameters $|\gamma| = 1.4 \times 10^{11}$ (rad/s) per T, $\mu_0 M_s = 18.9$ mT, $\varDelta = 7.4 \times 10^{-8}$ m. From equation 4.88 it follows that $v_W = 98$ m/s. Vella-Coleiro (1976c) observed initial wall velocities of 80, 93 and 95 m/s under drives of 1.3, 2.4 and 4.4 mT before some dramatic change took place in the wall structure and the velocity fell. This happened within 5 or 10 ns. Another report of velocities close to v_W was made by Telesnin *et al.* (1977) who observed a maximum wall velocity of 45 m/s in a bubble film for which $v_W = 41$ m/s.

We may now consider what must happen as the motion illustrated in figure 4.22b breaks down because the drive field is too high. The answer is shown in figure 4.23. In figure 4.23a we see the situation which must apply just before motion at the peak velocity occurs. At both the top and the bottom of the film the magnetisation cants out of the wall plane, against the direction of the local magnetostatic in-plane field, and the angle of canting is around the 30° which figure 3.7 would suggest for the case of $|B_y| \approx \mu_0 M_s$ and $Q \gg 1$. This means that

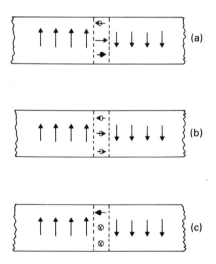

Figure 4.23 As the drive field is increased, beyond that which applies for figure 4.22b, the magnetisation must leave the wall plane, starting at the bottom surface as in (a) for the particular choice of initial screw-sense and sign of B_z made here. The wall screw-sense then reverses completely, (b) to (c)

the top part of the moving wall in figure 4.23a has an anticlockwise screw-sense while the bottom half has a clockwise screw-sense. Near the centre of the wall we find a Néel structure.

The exchange field, which arises from the twist in the wall structure throughout the thickness of the film, is helping to stabilise the dynamic wall structure at the top surface shown in figure 4.23a but is in opposition to the stabilising magnetostatic field at the bottom surface. It follows that the dynamic wall structure first collapses at the bottom surface of the film, the twisted wall structure unwinds, figure 4.23b, the stabilising effect of the exchange field at the top of the film is then lost and the continuous rotation of M, in and out of the wall plane as the wall moves forward with a fluctuating velocity, may begin. The mean value of this fluctuating velocity should be the saturation velocity discussed in section 3.9 of chapter 3 and given by equation 3.27.

Some quantitative detail of this very complicated process for straight walls may be found, from different points of view, in the theoretical work of Slonczewski (1973), Hubert (1975) Nedlin and Shapiro (1975) and Schlömann (1976). Our aim here is to move at once towards the far more practical case of the bubble domain in translation where the picture presented in figures 4.22 and 4.23 immediately suggest some very interesting complications.

Figure 4.24 shows a section through the simplest of all the bubble states which have been considered in previous sections: the unichiral $S = +1$ state. Applying the argument of figure 4.22 to this bubble when it is in translation makes it clear that the initial state, figure 4.24a, must develop into the dynamic state shown in figure 4.24b when a gradient drive field, ΔB_z is applied. This means that, as the drive field upon the bubble in translation is increased, the structure of the leading wall must begin to break up at the bottom surface while the trailing wall will begin to break up at the top.

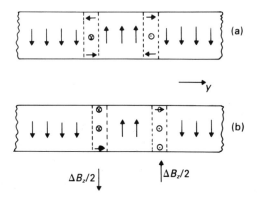

Figure 4.24 The initial wall structure for a bubble domain of the $S = +1$ state is shown in cross-section in (a). When a gradient drive is applied, ΔB_z across the bubble diameter, the leading wall begins to unwind at the bottom while the trailing wall begins to unwind at the top

A very complicated structure must develop within the closed bubble wall and this is best illustrated by means of a development of the wall structure, as Hagedorn (1974) does. This is shown in figure 4.25.

Figure 4.25a shows the wall structure of the unichiral $S = +1$ bubble viewed from the inside of the bubble and unwound to give a developed view. The very

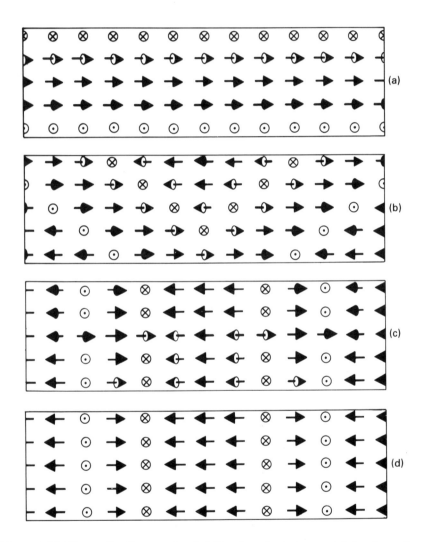

Figure 4.25 The wall of a moving bubble domain is unwound to show the changing structure as the bubble moves forward. Initially the state is $S = +1$, (a). Viewed from the inside, with the leading face of the bubble set in the centre of the diagram, (b), (c) and (d) show the development of a pair of Néel lines upon either side of the moving bubble

gentle twist in the wall structure, due to the magnetostatic field (figures 3.2 and 3.3) may be seen. This is the initial or static state.

Figure 4.25b shows the dynamic state which corresponds to the straight wall state first shown in figure 4.23a. This is the state just prior to the collapse of the high velocity mode of motion where M cants out of the wall plane against the stabilising influence of the magnetostatic field. The centre of the leading face of the moving bubble is in the centre of all the diagrams shown in figure 4.25.

Figure 4.25c shows the wall immediately after dynamic collapse. The magnetisation has rotated by one half revolution, right out of the wall plane and back again, at the bottom of the leading wall of the bubble, which is at the bottom centre of figure 4.25c, and at the top of the trailing wall where the cut has been made to produce the developments of the cylindrical wall structure shown in figure 4.25. It is immediately clear that this rotation of the wall M-field produces a twist around the circumference of the bubble which collects at the sides of the moving bubble domain. The final stage is shown in figure 4.25d; a pair of Néel lines has been generated upon either side of the bubble.

Figure 4.26 shows another view of what has happened. The state which has developed in figure 4.25d is now shown in figure 4.26b, looking down upon the moving bubble. Two points are made clear in this view: the bubble domain has a magnetostatic continuity, in the sense discussed in section 4.9 in connection with figure 4.7, and it still has the $S = +1$ state. The two Néel lines at the bottom of figure 4.26b contribute -1 to the state number while the two Néel lines at the top of the diagram contribute $+1$. The net contribution from all the Néel lines is zero.

The next stage in what Vella-Coleiro (1973) calls 'dynamic conversion' is shown in figure 4.26c. A further cycle of the process shown in figures 4.25b to d has taken place and the moving bubble now has eight Néel lines, four on either side. The mobility of this bubble is being reduced, step by step, as more Néel lines are generated and the additional viscous drag, which is associated with these

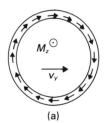 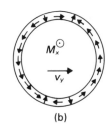 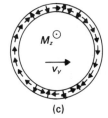

(a)   (b)   (c)

Figure 4.26 **Looking down upon the top of the moving bubble (a) corresponds to figure 4.25a while (b) corresponds to figure 4.25d. The last diagram, (c), shows the structure after yet another cycle of Néel line generation**

lines, is added to the drag upon the moving bubble. The state remains $S = +1$, however.

Before considering some of the implications of dynamic conversion some attempt must be made to add a quantitative aspect to what has so far, and understandably in view of the complexity of the problem, been a very qualitative description of what happens to a bubble domain when translated under high drive. This can be done by using the data given by Vella-Coleiro (1976a) who observed high drive bubble translation in a LuGdAlFe garnet layer, 9.4 μm thick. The bubble diameter was 6.1 μm and $\mu_0 M_s$ was given as 20.2 mT. Under a drive, ΔB_z of figure 4.24, of 0.93 mT the bubble was observed to move at a velocity of 42 m/s for the first 100 ns and then slow down as though dynamic conversion had taken place.

Under such high drive we may neglect the losses and ask if the total energy input to the bubble during time τ

$$E_{IN} \approx (\pi R^2 h)(2M_s)(\Delta B_z/2R)v_y\tau \tag{4.89}$$

is of the same order as the additional exchange energy stored in the Néel line pairs shown in figure 4.26b. This exchange energy is easy to estimate because we may argue that there will be $k(T_c-T)$ J associated with each pair of magnetic unit cells which are effectively rotated by a complete winding of the exchange link in going from figure 4.26a to 4.26b. The number involved is clearly twice the number which would fit into an area of dimensions wall width x bubble height, in this case approximately 10^{-7} m x 10^{-5} m. With a unit magnetic cell of dimensions 10^{-9} m the number involved is 2×10^6.

We are thus concerned with a comparison of E_{IN}, given by equation 4.89 and $2k(T_c-T) \times 10^6$. Using the data from Vella-Coleiro (1976a) quoted above we find $E_{IN} \approx 2 \times 10^{-15}$ J. The value of T_c-T for this garnet is about 100 K so that, rather fortuitously, $2k(T_c-T) \times 10^6$ is also approximately 2×10^{-15} J. It is clear that the 100 ns, observed by Vella-Coleiro to be the time required before dynamic conversion takes place, fits in with the mechanism proposed to explain the changes which must occur in the wall to produce the observed drop in the bubble velocity.

Some of the implications of dynamic conversion may now be discussed. Let us first consider what it may mean for a bubble domain device because the drive field is effectively on continuously in such a device and the Néel lines may build up on either side of the bubble, as shown in figure 4.26, never get the opportunity to meet and annihilate one another. The bubble mobility will gradually be degraded. This may have very serious consequences, as demonstrated by Shumate and Pierce (1973) who measured the gradual loss of data stored in a simple loop of a bubble memory as this data was continuously recycled. This loss of data was not found when a high-g garnet was used in the device (LeCraw et al., 1975) and this brings us to our final point because it was mentioned in section 1.10, that bubble overshoot was not found in high-g garnets either.

Ballistic overshoot in bubble domain translation is perhaps the most dramatic of all the effects in bubble dynamics which have been revealed by high speed microphotography and this was discussed from the experimentalist's point of view in section 1.10. The reason for bubble overshoot was first discussed at length by Malozemoff and Slonczewski (1975) and their model was elaborated and modified in later papers by Malozemoff and DeLuca (1978) and Slonczewski and Malozemoff (1978). It is clear from figure 4.26c that the bubble domain has a considerable amount of stored energy, after dynamic conversion, in the form of the Néel line pairs upon its sides. Hagedorn (1974), Henry *et al*. (1976) and Malozemoff and Slonczewski (1975) discussed the possibility of storing even more energy by the argument that an ion-implanted layer, or a multilayered garnet, would not allow M to rotate completely at the surface, as we have allowed here in going from figure 4.25b to 4.25c, so that dynamic conversion would produce not only Néel line pairs at the sides of the bubble domain but a tightly wound M field at the interface between the bubble film and ion-implanted or grown capping layer. There is certainly an important point here in view of the very large overshoot effect observed in the triple layer films used by Henry *et al*. (1976) and in view of the interesting asymmetric dynamic effects observed by Hannon (1978).

It remains to be explained how the energy which is stored in the twisted M field of a bubble, that has undergone dynamic conversion, is released in such a way as to propel the bubble in the same direction as it was moving in the first place. Malozemoff and DeLuca (1978) give an explanation in terms of the general theory of bubble dynamics which was described in section 4.10.6. This general theory attributes a sideways force upon the Néel lines which are clustered at the sides of the moving bubble, figures 4.26b and c, of such a sign that the bubble becomes elliptically distorted, expanding in a direction perpendicular to the direction of motion. When the pulse terminates, the bubble relaxes back and the radial motion of the Néel lines now produces a force which pulls the bubble forward. Good agreement with this theoretical model was claimed by DeLuca *et al*. (1977b) and by Malozemoff and DeLuca (1978). The model explains the overshoot without recovering the energy stored in the Néel lines, however, and this is rather unsatisfactory. There are still problems with the experiments in any case because DeLuca *et al*. (1977b) saw no rapid initial motion, in contrast to Vella-Coleiro (1976a), while Vella-Coleiro *et al*. (1975) saw no elliptical distortion of the bubble domain under high drive translation.

From the point of view which has been taken here it seems more likely that the viscous drag upon the Néel line clusters, of the kind shown in figures 4.26b and c, must play an important part in this problem. This viscous drag would cause the Néel lines to lag begind the diametrically opposed positions shown in figures 4.26b and c so that when the drive pulse ended the leading part of the bubble wall would dominate the trailing part magnetostatically, simply because it would be larger, and the Néel line pairs would be driven backwards to be annihilated. This would produce forward motion of the bubble itself. Clearly, more experimental work is needed before the mechanism behind bubble over-shoot will become really clear but high speed microphotography of normal or soft bubbles in translation is very difficult. High velocities are involved, DeLuca and Malozemoff (1976) have worked with velocities as high as 120 m/s, and the events are truly single shot events when dynamic conversion is involved: the final state of the bubble need not be the same as its initial state and a repetitive experiment is not possible.

5 Ferromagnetodynamics of Conducting Media

5.1 Introduction

The very early experimental work in ferromagnetodynamics, which was described in sections 1.2 and 1.3, was all done using conducting magnetic materials and was more concerned with the measurement of material properties, such as eddy-current loss and permeability, than with what was then the very new concept of domain wall motion. A valuable review of the beginnings of a more microscopic approach to these problems has been given by Kittel (1946a) who considered the motion of a domain wall separating head-to-head domains in a conducting magnetic material. This rather unexpected relative orientation of adjacent domains was criticised by Néel (1951) who considered what appears to be quite natural today, the 180° domain wall motion which has been such a dominant feature of previous chapters here. The results of these calculations by Néel (1951) were still expressed by the dispersion or frequency dependence of the real and imaginary part of the permeability of the material, however, and this point of view persists in many interesting and more recent papers by Lee (1958, 1960), Bishop and Lee (1963), Mulhall (1964), Zhakov and Filipov (1974) and Filipov and Zhakov (1975) who all give good reviews of the literature.

The idea of an equation of motion for the magnetisation in a conducting magnetic material and of directing attention towards the velocity and position of a domain wall really started with the publication by Stewart (1950) and by Williams et al. (1950) of their work on single crystals of silicon iron. This work is described in the next section as an introduction to the problem of ferromagneto-dynamics of conducting media.

5.2 Simple Domain Wall Motion in a Thin Sheet

The problem considered by Williams et al. (1950) was the motion of a single isolated domain wall separating two domains of the kind shown in figure 5.1 in a single crystal of 3% SiFe. Stewart (1950) performed the same experiment independently but his results involved an unknown integer factor because the number of walls involved was not known. The single wall experiment has been repeated under the most well defined conditions by Hellmiss and Storm (1974) using 3.5% SiFe.

As shown in figure 5.1a, the moving domain wall is plane. The implications of this assumption were discussed in section 2.18, and it was Williams et al. (1950) who were perhaps the first to consider the possibility that the moving wall might become curved under too large a drive field. This problem was discussed in

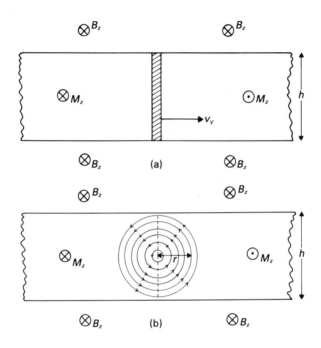

Figure 5.1 A thin sheet of conducting ferromagnetic material, lying in (y, z), has a plane domain wall, (x, z), moving under the influence of the drive field B_z with velocity v_y, as shown in (a). The motion of the wall sets up the circulating eddy-currents shown in (b)

section 2.18 in terms of the very small characteristic length, l, of nickel ferrite and YIG, the materials considered in chapter 2, their properties being summarised in table 2.1. A similar table is given here for the alloys which will be dealt with and the drive field, equation 2.119

$$\hat{B} = \mu_0 M_s \, l/h \qquad (5.1)$$

above which wall curvature might occur, is given in table 5.1 for the values of sample thickness, h, which were used in the experiments to be described. Despite the fact that some of the values of $\hat{B}$ given in table 5.1 are very small indeed we shall be able to justify the assumption that the effective drive field at the wall is well below $\hat{B}$, and that wall bending does not occur in the low drive field experiments which are the subject of this section.

Figure 5.1b shows the eddy-currents which must develop around the moving wall when the material has a finite conductivity. The expression 'eddy-current' is a good one because we should look upon these closed currents as being left behind in the material after the wall has passed by. For the rather thin samples and low velocities which we shall consider at first the relaxation time of the

Table 5.1 The saturation magnetic field, $\mu_0 M_s$ and conductivity, σ of the materials which will be discussed in this section are given. These are followed by approximate values for exchange constant, A, and anisotropy constant K. Parameters Q, Δ and ℓ are then calculated. The thickness, h, actually used in the experiment is then given after which $\hat{B}$ and the eddy-current limited mobility from equation 5.8 may be calculated. Data is taken from the references cited in figure 5.2 or from Bozorth (1951)

	Si_3Fe_{97}	$Ni_{80}Fe_{20}$	$Ni_{45}Fe_{30}Co_{25}$	$Ni_{65}Fe_{35}$	$Fe_{40}Ni_{40}P_{14}B_6$	Unit
$\mu_0 M_s$	2.02	1.00	1.50	1.30	0.84	T
σ	2.5×10^6	5×10^6	5.4×10^6	4.8×10^6	6.2×10^5	mho/m
A	10^{-11}	10^{-11}	10^{-11}	10^{-11}	10^{-11}	J/m
K	3×10^4	1.5×10^2	2×10^2	5×10^2	1.4×10^3	J/m^3
$Q = 2K/\mu_0 M_s^2$	2×10^{-2}	4×10^{-2}	2×10^{-4}	7×10^{-4}	5×10^{-3}	–
$\Delta = (A/K)^{\frac{1}{2}}$	2×10^{-8}	2.5×10^{-7}	2×10^{-7}	1.5×10^{-7}	10^{-7}	m
$l = 2Q\Delta$	8×10^{-10}	2×10^{-10}	8×10^{-11}	2×10^{-10}	10^{-9}	m
h	1.14×10^{-3}	0.2×10^{-6}	1.47×10^{-4}	1.25×10^{-4}	5×10^{-5}	m
$\hat{B} = \mu_0 M_s l/h$	1.4×10^{-6}	10^{-3}	8.2×10^{-7}	2.1×10^{-6}	1.7×10^{-5}	T
$\mu_w = 4/\mu_0^2 \sigma h M_s$	5.6×10^2	3.6×10^6	2.7×10^3	4.1×10^3	1.2×10^5	m/s per T

eddy-currents, and this will be shown to be of the order of $\mu_0 \sigma h^2$, is so small that the eddy-current pattern appears to move along as though it was attached to the moving wall. Nevertheless, in order to calculate the magnitude and sign of the currents shown in figure 5.1b we should apply the Maxwell integral equation

$$\oint E \cdot dl = -\frac{d}{dt} \iint B \cdot ds \qquad (5.2)$$

to a closed contour in the stationary medium and consider the instant when the domain wall bisects this closed contour. Taking the circular contours shown in figure 5.1b, equation 5.2 is

$$2\pi r E_\theta = -(2\mu_0 M_s)(2r \, dy/dt) \qquad (5.3)$$

where r is the radius of the contour. This simple model leads to a value of E_θ which is independent of r and a current density

$$J_\theta = -2\mu_0 \sigma M_s v_y/\pi \qquad (5.4)$$

The total current flowing around the domain wall per unit length is then $hJ_\theta/2$ and this current will produce a magnetic field at the centre of the wall region given by

$$(B_e)_{r=0} = -\mu_0 \sigma h M_s v_y/\pi \qquad (5.5)$$

We now follow an argument given by Carr (1976a) which will be discussed in more detail in the next section. This argument is that the total magnetic field needed to drive the wall, in the absence of the eddy-current damping, is vanishingly small, because it is only needed to overcome the very small Landau–Lifshitz damping mechanism discussed in chapter 2. Carr (1976a) concluded that the drive field, B_z, in which the sample is shown to be immersed in figure 5.1a is only needed to overcome the eddy-current field, given by equation 5.5. It follows that we need only set

$$B_z + (B_e)_{r=0} = 0 \qquad (5.6)$$

combine equations 5.5 and 5.6 and solve for v_y in terms of B_z

$$v_y = (\pi/\mu_0^2 \sigma h M_s) B_z \qquad (5.7)$$

to obtain an estimate of the eddy-current limited mobility.

This very simple argument has been given here to bring out the essential facts. A detailed treatment of the problem requires that we find the mean value of B_e over the entire wall region and balance this with B_z. This is quite an involved, but very straightforward, calculation and has been given by Williams *et al.* (1950) Patton *et al.* (1966) and Carr (1976a) who all find that the velocity is given by the expression in equation 5.7 except that the constant, π, should be replaced by a number close to 4. The model itself is not accurate enough for some of the numerical detail which may be given to it, however, and we shall use the expression

$$\mu_{w} = (4/\mu_{0}^{2} \sigma h M_{s}) \qquad (5.8)$$

as our final result for the eddy-current limited mobility of domain walls in cases of the kind illustrated in figure 5.1. Because this result has been obtained by balancing the fields B_{z} and B_{e}, along the lines of equation 5.6, it is clear that we have justified the assumption that the effective drive field at the wall is much smaller than B_{z} which means that B_{z} may exceed the value of $\hat{B}$, given in table 5.1, by at least an order of magnitude and wall bending will still not occur.

Figure 5.2 shows how well this model fits the experimental facts. By plotting μ_{w}, from equation 5.8, as a function of sample thickness, h, we get a number of

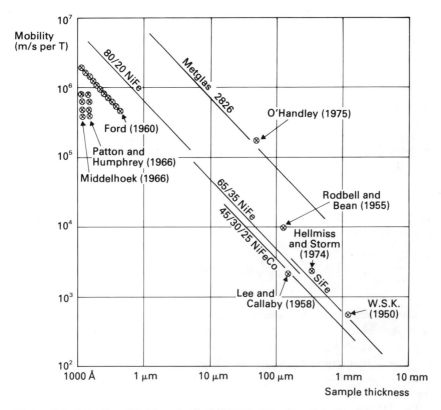

Figure 5.2 Equation 5.8 is compared with the experimental data from a number of sources. W.S.K. is Williams, Shockley and Kittel (1950)

straight lines on a log–log plot, all of slope −1 and each representing one of the materials listed in table 5.1. The available experimental data covers a very wide range of sample thickness and it is clear from figure 5.2 that the measured low drive field mobilities are all in good agreement with equation 5.8 except in the case of the very thin permalloy films, studied by Ford (1960), Middelhoek (1966) and Patton and Humphrey (1966), which will be considered in section 5.4. O'Handley (1975) gave results for 50 μm thick Metglas ribbon, which were made using very low drive fields and should not be confused with his experiments on Metglas wires which were discussed in chapter 1, figure 1.3. The results given by Lee and Callaby (1958), discussed in section 1.5, are seen to be in excellent agreement with equation 5.8 for the parameters of their material, 45/30/25 NiFeCo or Perminvar. Figure 5.2 shows similar excellent agreement for the measurements of Williams *et al.* (1950) on 3% SiFe and the results given by Hellmiss and Storm (1974) for 3.5% SiFe, 350 μm thick, are also shown in figure 5.2 to be very close to the SiFe straight line, which has been drawn for the 3% Si material but the difference between the two alloys is negligible. Rodbell and Bean (1955) used an indirect method which might have involved more than a single domain wall and this would explain why their result lies above the line for 65/35 NiFe. The results given by Konishi *et al.* (1970) for 50/50 NiFe tapes are not shown in figure 5.2 for the same reason; more than one wall was involved so that the apparent mobility was much higher than equation 5.8 would predict.

We may now continue and ask why the low drive field mobility of the very thin permalloy films is so much lower than the value expected from equation 5.8, as shown on the far left of figure 5.2. Before doing this, however we should consider some fundamental differences between the kind of equation of motion we are using here, for the ferromagnetodynamics of conducting media, and the equation of motion used in the previous chapters for insulating materials.

5.3 The Equation of Motion for Conducting Media

The ferromagnetodynamics of conducting media is governed by the same Landau–Lifshitz equation of motion

$$\frac{-1}{|\gamma|} \frac{\mathrm{d}M}{\mathrm{d}t} = (M \times F) - \lambda \left[F - (F \cdot M) M/M_s^2\right] \tag{5.9}$$

which has been used in the previous chapters but the total effective field, F, is now given by the sum of five distinct terms. These are the four terms which went to make up F when it was first introduced here, in chapter 2, equation 2.2, the contributions from exchange, anisotropy, the magnetostatic field and the externally applied field, plus a fifth term or part of F which is the field due to the eddy-currents.

Carr (1976b) has made the important point that the eddy-current contribution to F cannot be treated as a local term in equation 5.9, as can the exchange, anisotropy and external field terms, but presents the same kind of difficulties as the magnetostatic field. The magnetic field which is produced at any point on the moving domain wall by the eddy-currents depends upon the total distribution of these currents throughout the entire sample. Just the same applied for the magnetostatic field which was introduced in section 2.5, and was found to depend upon the distribution of div M throughout the entire sample.

To express the equation of motion more formally we take the Maxwell equations

$$\text{curl } E = -\partial B/\partial t \tag{5.10}$$

and

$$\text{curl } H = J + \partial D/\partial t \tag{5.11}$$

and neglect the displacement current $\partial D/\partial t$ compared to the conduction current

$$J = \sigma E \tag{5.12}$$

It is important to note that we assume a scalar conductivity here. This is rather a serious simplification in a magnetic material and has some interesting implications which will be discussed in section 5.9.

We now substitute equation 5.12 and the general relationship

$$B = \mu_0 (H + M) \tag{5.13}$$

into equations 5.10 and 5.11, combine these and obtain

$$\nabla^2 J = \mu_0 \sigma \frac{\partial}{\partial t} (J + \text{curl } M) \tag{5.14}$$

as the equation which relates the eddy-curents in the medium to the time and space derivatives of M.

In the particularly simple cases where M and all its derivatives are known, equation 5.14 may be solved for J and then the magnetic field due to the eddy-currents calculated using the well known Biot–Savart relationship

$$B_e = \frac{\mu_0}{4\pi} \int \frac{(J \times r)}{r^3} \, dv \tag{5.15}$$

(Sommerfeld, 1952). In equation 5.15, the integral is taken over the entire volume occupied by J and r is the vector joining the point of integration to the point where B_e is to be evaluated. The result may be substituted into equation

5.9 as the eddy-current contribution to F provided the problem allows simple superposition, but this is really never the true situation in a ferromagnet and, to return to our primary assumption above, cases in which M and all its derivatives are known are cases which are already solved. It follows that there are no simple solutions to problems in the ferromagnetodynamics of conducting media, just as there are no simple solutions to the problems which were discussed in the previous chapters when the magnetostatic field had to be taken into account. An accurate solution may only be attempted numerically by assuming some initial dynamic distribution of M, calculating the eddy-currents and the magnetostatic field, calculating the eddy-current field and then modifying the initial distribution of M which was assumed in the hope of eventually converging to a stable solution.

Some very interesting distortions of the moving domain wall may come about because of the eddy-current magnetic field. This kind of problem, and its growing literature, will be discussed briefly in section 5.5 but the essential details may be illustrated with the aid of the analysis which was given in the previous section and by reference to figure 5.1b. If we continue to adopt the model that the eddy-current field is balancing out the applied drive field, B_z, it is clear from figure 5.1b that this balance is only possible well inside the sample and that at the top and bottom surfaces of the sample the moving wall will see almost all of the drive field. It follows that these parts of the wall should move forward more rapidly than the central parts. This wall bending would be resisted by the wall surface energy in just the same way as wall bending of the opposite sense was resisted in the insulating magnetic materials discussed in section 2.20.

Eddy-currents are not the only currents which may flow in a conducting medium and we should really have left the fifth contribution to F, discussed above, more general and said that it was due to currents flowing in the material. These currents may be supplied from an external source and we are brought back to the remarks concerning the conductivity, σ, made in connection with equation 5.12. In a ferromagnet, which has domains, a current which is supplied from outside may take up a very complicated distribution inside the sample because of strictly magnetic effects such as the Hall effect or the anisotropy of magnetoresistance. Complex effects of this kind have been studied and will be discussed in section 5.9.

Finally, there is an alternative approach to the problem of domain wall motion in conducting media. This is to argue that if the eddy-currents are the main kind of loss in a problem, so that $\lambda \approx 0$ in equation 5.9, the total dissipation, $B \cdot J$ or σE^2, should be calculated and equated to the work done on the moving wall by the external field. This was, in fact, the method adopted by Williams *et al.* (1950) in their derivation of equation 5.8. Carr (1976a) has contrasted this point of view with the detailed field model discussed previously and concludes that the method which involves a calculation of the losses is only useful when the moving wall may be assumed to be rigid. Nevertheless, a solution to the problem of some kinds of wall motion is made very much easier when the dissipation or Joule heating, σE^2, point of view is taken and an example is given in the next section.

5.4 The Low Drive Mobility in Thin Permalloy Films

Figure 5.2 shows that the low drive field wall mobility measured in very thin permalloy films is well below the mobility predicted by equation 5.8. There are a number of very important points about these experimental facts.

In the first place, we are concerned here with wall mobility. The vast literature on thin permalloy films, that is films of the very low anisotropy $Ni_{80}Fe_{20}$ alloy, which came out between 1957 and 1967 is, in fact, more concerned with switching the direction of the magnetisation in a thin film. This is the subject of section 5.7. Experimental work which concerns itself with single isolated walls, using magneto-optical techniques to ensure that a single wall is being observed, is rare in comparison and we can add only a few later references to the three already cited in figure 5.2 — these are Telesnin *et al.* (1969), Konishi *et al.* (1971), Bartran and Bourne (1972, 1973) and Kolotov *et al.* (1975). The first three of these obtained very good agreement with the earlier measurements of Patton and Humphrey (1966) and Middelhoek (1966) in that the low drive field mobilities they measured lay between 5×10^5 and 10^6 m/s per T for films of thickness between 0.1 μm and 0.3 μm. The thinner films showed higher mobilities and some experimentalists observed a strong minimum in the mobility around a thickness of 0.07 μm. This minimum will be discussed later.

Secondly, we are concerned with the low drive field regime. In the previous sections of this chapter, concerned with wall motion in quite thick samples, low drive field has meant keeping the effective drive field well below $\hat{B}$, the field given in table 5.1 above which wall bending may occur. There is clearly no problem from this point of view with the thin $Ni_{80}Fe_{20}$ films because $\hat{B} \approx 1$ mT. The problem with the thin films is that the absolute value of the wall velocity may begin to approach the velocity $|\gamma|(\mu_0 A)^{1/2}$, discussed in section 2.17, and equal to about 500 m/s for the metals, where changes in even the simplest wall structures would be expected. The wall velocities in these thin films can be much higher than in the previous examples. Bartran and Bourne (1973) observed velocities up to 150 m/s, for example.

Fortunately, changes in wall structure are fairly easy to identify experimentally as a discontinuous drop in the wall mobility. Patton and Humphrey (1966) observed this at about 35 m/s in their films and some data on this kind of behaviour has been collected by Patton (1973). It seems safe to say that wall velocities below 10 m/s belong to the low drive field regime in these thin films.

A third point to be taken into consideration when discussing wall mobility in thin permalloy films is that the coercivity is very much higher than it is for the much thicker samples included in figure 5.2. As Hoffman (1973) does, we must clearly distinguish three different coercivities in thin films.

(1) The nucleation field, which is necessary to nucleate reverse domains after a previous saturation.

(2) The growth field, which is necessary to cause a nucleated domain to grow by domain tip motion.

(3) The wall coercivity, which is the field necessary to move an existing wall.

In the experiments under discussion, it is the third kind of coercivity, the wall coercivity, which is of importance. This may be as high as 0.1 mT in a thin permalloy film because it depends upon the polycrystalline nature of these films, the fact that their properties may show spatial dispersion and that there may be quite gross defects in the form of sub-micron dust particles on the substrate, or pin-holes. It is very important to realise that this coercive field must be exceeded, by definition, before any wall motion can be recorded by the before and after Kerr effect observations used by all the experimentalists cited except Ford (1960), who used the Kerr effect to actually measure the wall velocity.

If we observe the initial position of a domain wall using the Kerr effect, pulse the drive field and then observe the final position we cannot say that the wall velocity is the distance moved divided by the pulse duration unless we are certain that the wall stops moving immediately the drive pulse ends. Exactly the same problem arose in connection with the bubble translation experiment discussed in chapters 1 and 4. In the case of the permalloy films, however, it is not overshoot which will cause error but the fact that the wall may move backwards once the drive pulse is terminated. This follows because gross defects, such as inclusions and pin-holes, will determine a whole series of stopping places for the wall as it moves across the film and in between these defects the wall must bend in the plane of the film. Ford (1960) would have measured a fairly accurate wall velocity as the wall crossed the illuminated region used in his experiments but the other experimentalists would have deduced a lower velocity than the true velocity because the wall would have moved back to being quite straight at the end of the experiment. This conclusion is supported by the most recent measurements of wall velocity by Kolotov *et al.* (1975) and by Kolotov and Lobachev (1975) who used high speed microscopy with a time resolution of 15 ns to observe the detailed motion of 180° domain walls in thin films of 4/79/17 Mo/Ni/Fe and 83/17 Ni/Fe. The true velocity of the wall was observed to be between 1.5 to 5 times higher than the velocity deduced from measuring the initial and final positions and then dividing by the drive pulse length. This was only true in the thickness region close to 0.07 µm, however, where all the previous investigators, except Ford (1960), had deduced a minimum in the mobility.

A final point must be mentioned concerning experiments which seek to measure the low drive field mobility in thin films and this again involves coercivity. When the wall coercivity is very small it seems quite acceptable to write

$$v_y = \mu_w \, (B_z - B_c) \tag{5.16}$$

as though the coercive field, B_c, simply subtracts from the drive field, B_z. In the thin film experiments, however, B_z may exceed B_c by only a few per cent and the wall velocity can already be close to the upper limit we might choose to define for low-drive conditions. Baldwin (1972) deals with this kind of problem in an important paper which is concerned with the finite hysteresis loss observed at zero frequency in magnetic materials and the situation is as follows. If the wall moves through a series of defects which cause parts of it to stop moving while other parts continue, that is the wall bends, the average velocity of the wall must be below the instantaneous velocity of some parts of the wall. The dissipation due to the wall motion depends upon v^2, however, so that this kind of motion must involve an additional loss when it is compared to absolutely uniform motion with the entire wall moving at the same constant speed.

Multiplying equation 5.16 by $2M_s v_y$ and rearranging we may write

$$2M_s B_z v_y = (2M_s/\mu_w)v_y{}^2 + 2M_s B_c v_y \qquad (5.17)$$

to show that the power input per unit area to the wall, $2M_s B_z v_y$, must equal the viscous dissipation which we would attribute to motion at a constant velocity, v_y, plus an additional loss which involves B_c as it is defined by equation 5.16. Baldwin (1972) shows that this B_c, as defined by the additional losses associated with the defects, is identical to the wall coercivity as defined by equation 5.16.

A very important conclusion from this is that if measurements of velocity against drive field are found to follow a relationship like equation 5.16 but B_c is not identical to the wall coercivity, as defined by Hoffmann (1973), then the experiment does not involve simple single wall motion at all or the moving wall must have a completely different structure to the static wall upon which the measurement of coercivity was made. Examples of this kind of behaviour will be given later in this chapter and have already been discussed in chapter 1 in connection with figures 1.2, 1.7 and 1.8.

We now return to the main problem of this section which is the low mobility of the thin $Ni_{80}Fe_{20}$ films when compared with the mobility predicted by equation 5.8. Patton and Humphrey (1966, 1968) attributed this to the intrinsic ferromagnetic damping, or Landau–Lifshitz damping, of the material but this meant that a value of $\alpha = \lambda/M_s = 0.014$ (equation 5.9 and section 2.12) had to be used and such a large value of α was compatible with the FMR measurements which had been made on thin films of this kind at the time (Hearn, 1964). Spin-wave resonance in thin metal films was a new field at that time, however, and has been reviewed by Weber (1968) and by Patton (1968), whose paper is particularly interesting for the references it gives to the part played by eddy-currents in resonance measurements. One of the main problems is that it may not be possible to excite a uniform precession mode of FMR in a thin film and make a direct measurement of α, in the way a spherical sample may be dealt with as described in section 2.12. If such an experiment is done, a very broad resonance may be observed, and thus a large value of α attributed to the film, but what is really being observed is the envelope of a poorly excited spin-wave spectrum. If the experiment is done at a low enough frequency, however, these problems can be avoided and an early example of this is the work by Rossing (1963) in the 100 MHz to 1.6 GHz range. This showed that there was a large contribution to the resonance damping coming from the spatial dispersion of the magnetic anisotropy which would certainly explain why many experimentalists had observed such wide line widths in permalloy because all samples were known to show anisotropy dispersion. More recent work by Patton *et al.* (1975) collects the data for resonance in permalloy films over the remarkably wide frequency range of 2 to 36 GHz.

All this work on resonance in thin permalloy films shows that the intrinsic damping in this material is quite small and that some other explanation must be found for the low value of the wall mobility shown in figure 5.2.

Patton *et al.* (1966) considered an additional kind of Joule heating, brought about by the wall motion, which was not brought into the problem by Williams *et al.* (1950). This was the Joule heat which would be developed by a current induced along the direction of the moving wall and we take care not to call this an eddy-current, before even inquiring how such a current might come about, for two rather fundamental reasons. Firstly, the current will be found to be intimately connected with the motion of the wall, not produced by the passage of the wall leaving an eddy-current behind. Secondly, the current does not produce a magnetic field which is in simple opposition to the applied field, B_z, as was the case for the circulating current shown in figure 5.1b.

It becomes clear that a current may be induced along the direction of the wall, because of its motion, when we consider the detailed wall structure. It should be noted that in section 5.2 the moving wall was treated as though it had zero width; any internal structure of the wall region was simply neglected. To begin with, consider the case of a fairly thick film $h \gg \varDelta$, where $\varDelta$ is the wall width parameter $(A/K)^{1/2}$ listed in table 5.1 for the kind of materials under consideration. The magnetisation lies in the plane of the film within the two domains upon either side of the wall and when $h \gg \varDelta$ we may assume that a simple Landau–Lifshitz wall structure applies, as we have discussed before in chapter 1, figure 1.9, and chapter 2, figure 2.2. It then follows that we have a component of magnetisation

$$M_x = M_s \text{ sech } (y/\varDelta) \qquad (5.18)$$

normal to the plane of the film and that this will produce a magnetic field $\mu_0 M_x$ because the surface divergence of M_x may be neglected near the wall when $h \gg \varDelta$.

When the wall is moving at velocity v_y, the field $\mu_0 M_x$ will produce an electric field along the wall direction given by $E = v \times B$, that is $E_z = -\mu_0 v_y M_x$. The total Joule heating per unit length of wall then follows by using equation 5.18, integrating and multiplying by the film thickness, h, to obtain

$$W_j = 2\mu_0^2 h\sigma v_y^2 M_s^2 \varDelta \qquad (5.19)$$

Now the power dissipated in Joule heating, W_j, must be equal to the rate of work done on the moving wall by the externally applied field B_z. This power input to the wall is simply

$$W_{IN} = 2hv_yM_sB_z \qquad (5.20)$$

per unit length.

We now equate equations 5.19 and 5.20 and solve for v_y to obtain

$$v_y = (1/\mu_0^2 \sigma M_s \Delta) B_z \qquad (5.21)$$

Now equation 5.21 gives the wall mobility as though the z-component of induced current was the only damping factor. Similarly, equation 5.8 gave the mobility as though the circulating eddy-currents of figure 5.1b were the only factor involved. To obtain an expression for the mobility which takes both factors into account we remember that the viscous drag on the wall is given by $(2M_s/\mu_w)$ per unit area of wall so that if we have two separate loss mechanisms which give results $v_y = \mu_{w1}B_z$ and $v_y = \mu_{w2}B_z$, we find the total effective mobility, μ_w, by writing

$$(1/\mu_w) = (1/\mu_{w1}) + (1/\mu_{w2}) \qquad (5.22)$$

Combining the expressions given by equations 5.8 and 5.21 leads to a final result

$$\mu_w = \frac{4}{\mu_0^2 \sigma M_s \ (h + 4\Delta)} \qquad (5.23)$$

At first glance this looks like an encouraging result in the light of the experimental facts shown in figure 5.2. As table 5.1 shows, $\Delta \approx 0.2$ μm for permalloy so that we would expect the mobility to be halved at a sample thickness of 0.8 μm, relative to the value shown in figure 5.2, and then become constant at $(\mu_0^2 \sigma M_s \Delta)^{-1}$, as $h < 4\Delta$, which is a mobility of 8×10^5 m/s per T, very close indeed to the mobilities measured by Patton and Humphrey (1966) and Middelhoek (1966) who also observed very little thickness dependence.

Such a conclusion would be incorrect however because we cannot continue to assume that equation 5.18 applies when the film thickness, h, becomes comparable with Δ. Patton et al. (1966) corrected for the surface divergence of M_x in their analysis of the problem, that is the so-called demagnetising effects when $h \approx \Delta$ and the field produced by M_x is less than $\mu_0 M_x$, but they continued to assume the Landau–Lifshitz wall structure equation 5.19. This is not surprising because their work predated the important developments in the theory of domain wall structure for very thin films of this kind, which began with the work of LaBonte (1969) and Hubert (1969), recently reviewed by Brown (1978), and continued with the work of Hubert (1974) and Aharoni (1975a, 1975b, 1975c).

Figure 5.3 is an illustration of the kind of wall structure which must exist in a permalloy film $0.1 \mu m$ thick and is adapted from Aharoni (1975c). The transition region between the two domains is seen to occupy a distance roughly equal to the film thickness, as we would expect, Δ being $0.2 \mu m$ for permalloy, but the region of rotation of M in the sense of equation 5.18, or figure 1.4. now occurs over a curved surface lying to the right of figure 5.3 instead of over a plane wall.

The contrast in the magnetic fields associated with this kind of wall and the simple Landau–Lifshitz wall which would be found in a much thicker film is shown in figure 5.4. Only the component of B which lies in (x, y) is shown in both cases. It is obvious from figure 5.4 that there is less magnetostatic energy associated with the magnetic field shown in figure 5.4b than with that shown in figure 5.4a because the field is much weaker outside the sample. In fact, the earlier models which led to the kind of solution illustrated in figure 5.3 used the idea that the surface divergence of M should be zero but Aharoni (1975c) showed that a better energy minimum was obtained when a small surface divergence in M, shown on the right-hand side of figure 5.3, was allowed.

We may now consider the eddy-current loss associated with the motion of the

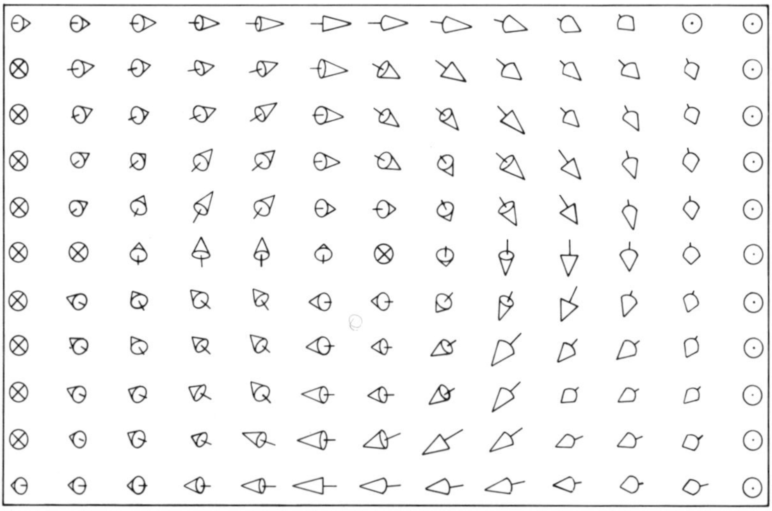

Figure 5.3 The M-field at the wall region in a permalloy film $0.1 \mu m$ thick. From Aharoni (1975c)

magnetic field shown in figure 5.4b through the conducting film as the wall moves along y. The maximum value of B_x will now be $\mu_0 M_s$ regardless of how thin the film may become because there will be no so-called demagnetising effects due to the surface divergence of M, as there are in the simple Landau–Lifshitz wall of figure 5.4a. It follows that there will be an electric field $E_z = \mu_0 v_y M_s$ close to the leading edge of the moving wall and a current $J_z = \sigma\mu_0 v_y M_s$. There will be a much weaker current associated with the trailing edge of the wall along $-z$. Note that there are four possible forms for this kind of wall: two for the choice of screw-sense, as for the Landau–Lifshitz wall, which determines whether the circulating field shown in figure 5.4b is clockwise or anticlockwise, and then both of these forms may have the region of high B_x either on the right-hand side, as in figure 5.4b or on the left-hand side.

This modification to the wall structure would make it possible that an expression for the low field mobility very much like equation 5.23 would, in fact, apply as the film thickness approaches 4Δ and the wall structure changes from a simple Landau–Lifshitz structure, equation 5.18, to the two-dimensional structure

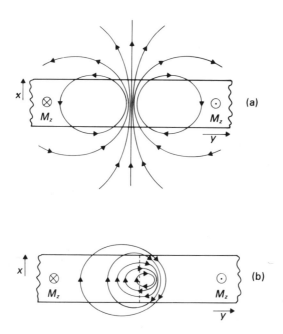

Figure 5.4 In (a), the *B*-field of a simple Landau–Lifshitz wall in a permalloy film is shown. The *B*-field from the wall shown in figure 5.3 is quite different and is shown in (b). (Copyright 1979, North Holland Publishing Co. Reproduced by permission from the Journal of Magnetism and Magnetic Materials, 13, 316)

illustrated in figure 5.3. If the film was made thinner still, that is $h < \Delta$, we would expect a further increase in mobility because the mobility given by equation 5.21 should now be the dominant factor but Δ in equation 5.21 should be replaced by h because the width of the kind of wall discussed by Aharoni (1975c) becomes more and more equal to h as the film is made still thinner. This means that we would replace equation 5.23 by

$$\mu_w = 1/(\mu_0^2 \sigma M_s h) \qquad (5.24)$$

and it is remarkable how well equation 5.24 agrees with the data given by Ford (1960), figure 5.2, who certainly used the more accurate experimental technique now that we may look back on this work in the light of the experiments of Kolotov *et al.* (1975) and of Kolotov and Lobachev (1975). Kolotov *et al.* (1978) have recently improved the space–time resolution of their system to 3 μm–5 ns.

Equation 5.24 is also in reasonable agreement with the high mobilities measured by Patton and Humphrey (1966), Middelhoek (1966), Telesnin *et al.* (1969) and Konishi *et al.* (1971) for films which were below 0.1 μm in thickness and not included in figure 5.2. This is a very interesting region and Aharoni (1976) has applied his work on the two-dimensional wall structures, of the kind illustrated in figure 5.3, to the problem of wall motion and has shown that these structures may not be subject to the kind of dramatic changes which have been attributed to the simpler one-dimensional structures (Schlömann, 1972, Bartran and Bourne, 1973) when the drive field is increased and the velocity begins to approach the limit discussed at the beginning of this section; $v \approx |\gamma|(\mu_0 A)^{1/2}$. Much interesting work is still to be done in this area (Aharoni, 1974, 1979, Aharoni and Jakubovics, 1979).

5.5 Domain Wall Motion in a Thin Sheet Under Conditions of High Drive

In the previous sections we considered domain wall motion in a conducting medium under steady state conditions. It was assumed that a domain wall was already present within the sample and that this wall was moving at a constant velocity. Such a simple situation does not really represent a practical problem, however, and we must ask how the dynamic process began, how it continued and how it finished. One of the very first attempts to look at the complete dynamic problem in this way was made by Williams *et al.* (1950) in their work on SiFe single crystal samples which was discussed in section 5.2.

Williams *et al.* (1950) considered what would happen in the case where their sample was magnetically saturated and then had a magnetic field applied to it which was intended to reverse the direction of the magnetisation so that the sample would end up saturated again but now in the opposite direction. Their model for what happens inside the sample under these conditions is shown in figure 5.5. The initially saturated sample of square cross-section is shown with reverse domains developing and then joining up to form a collapsing cylindrical

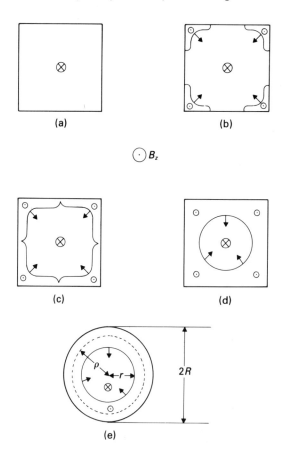

Figure 5.5 Showing the development of the reverse domains during magnetic reversal of a square conducting sample, (a)–(d). The idealised cylindrical model which is used is shown in (e)

domain in the centre of the sample. This model corresponds to the kind of experiment being performed by Williams *et al.* (1950) because their sample had an almost square cross-section, 1.14 mm x 1.52 mm.

In figure 5.5 the initial state of complete magnetic saturation, a, is shown to respond to the applied field, B_z, by the nucleation of reverse domains at its corners, b. The reason for this is that the applied field may more easily penetrate at a sharp corner like this because there is no way for the eddy-currents to close and contain the moving wall until there is one continuous domain wall around the entire sample, as in figure 5.5c. This contracting or collapsing domain wall would rapidly become cylindrical as shown in figure 5.5d.

The problem is modelled in a rather idealised way in figure 5.5e where a completely cylindrical geometry is assumed. The collapsing domain wall is at radius r and we use equation 5.2 to obtain a relationship between the θ-component of the electric field at radius ρ and the velocity of collapse. This is found to be

$$E_\theta = -2\mu_0 r M_s |v_r|/\rho \tag{5.25}$$

which drives a current around the outside of the wall in a clockwise direction to produce a magnetic field which is in opposition to B_z. As in section 5.2, we argue that this field must be equal and opposite to B_z.

The total current per unit length for the case shown in figure 5.5e is given by

$$I_\theta = -2\mu_0 \sigma r M_s |v_r| \int_r^R \frac{d\rho}{\rho} \tag{5.26}$$

and the field which must balance B_z is simply given by multiplying equation 5.26 by μ_0. Completing the integration and solving for v_r gives

$$|v_r| = B_z / [2\mu_0^2 \sigma M_s r \log_e (R/r)] \tag{5.27}$$

Equation 5.27 is clearly the result of an over-simplification because it predicts infinite initial velocity for the cylindrical model of figure 5.5e when $r = R$. Nevertheless, Williams *et al.* (1950) obtained excellent agreement between this model and their experiments on single crystal samples of SiFe using quite large drive fields, up to 8 mT. This was because their samples were quite large, so that magnetic reversal took place during several hundred microseconds and the details of the beginning of the process of reversal, figure 5.5a to c, were lost.

Experiments of this kind which did begin to show some of the more complicated behaviour expected in magnetic reversal were reported by Menyuk (1955) who studied the magnetic switching properties of very thin metal alloy tapes, wound in the form of toroidal cores which had been suggested as a possible alternative to ferrite cores in computer memories (Menyuk and Goodenough, 1955). Under high drive conditions these tapes would be expected to reverse their magnetisation in the manner shown in figure 5.6 and end up with a collapsing elliptical domain, as shown in figure 5.6b. Because the thickness to width ratio, h/W is very small in these tapes, typically 1/100, we can model this very simply, as shown in figure 5.6c.

Applying equation 5.2 to the situation shown in figure 5.6c, we find an electric field

$$E_x = 2\mu_0 M_s v_y \tag{5.28}$$

everywhere behind the moving wall, between $h/2$ and y, at the top of figure 5.6c and the same electric field, but with opposite sign, behind the moving wall at the bottom of figure 5.6c. A current density, $J_x = \pm \sigma E_x$, thus exists everywhere

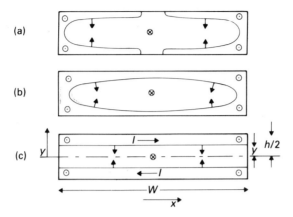

Figure 5.6 Showing the development of the reverse domains during magnetic reversal in a thin tape, (a)–(b). The idealised model is shown in (c)

in between the two moving walls and the top and bottom surfaces of the thin metal tape. The centre of the tape has no current flowing in it.

The total current per unit length is shown as I in figure 5.6c and will be given by

$$I = 2\sigma\mu_0 M_s v_y \ (h/2 - y) \tag{5.29}$$

which produces a magnetic field in opposition to the applied field, B_z, given by $\mu_0 I$. Using the hypothesis that this field must, in fact, equal B_z we may solve for v_y to obtain

$$v_y = [1/2\mu_0^2 \sigma M_s \ (h/2 - y)] \ B_z \tag{5.30}$$

which, like equation 5.27, is rather unsatisfactory in that v_y is infinite when $y = h/2$ but, as figure 5.6a shows, the beginning of the magnetic reversal is not covered by this model; it is only applicable when the reversal process is well under way.

Since the development of magnetic core memories (Forrester, 1951) it had become established practice to characterise the cores by means of a switching coefficient, S_w, defined by the equation

$$\tau^{-1} = (S_w)^{-1} \ (B - B_0) \tag{5.31}$$

where τ is the magnetic reversal time, B is the applied drive field and B_0 is a constant of the material. A linear relationship between τ^{-1} and the drive field is, in fact, seen in many of the ferrite materials which were used in computer memories and was also seen by Menyuk (1955) in the thin metal tape cores he studied. A very rough estimate of the switching coefficient may be made by arguing that the reversal time will be made up from a thickness independent part, inversely proportional to the applied field and representing the process illustrated in figures 5.5a to 5.5b, followed by a part which is due to wall motion through the thickness of the tape, as illustrated in figure 5.5c, and governed by equation 5.30. Taking an average velocity for this latter part, by setting $y = h/4$ in equation 5.30, this means that we have a reversal time given by

$$\tau = (K + \mu_0^2 \, \sigma M_s h^2/4) \, B_z^{-1} \qquad (5.32)$$

and

$$S_w = (K + \mu_0^2 \, \sigma M_s h^2/4) \qquad (5.33)$$

where K is a constant. This model may be compared with experiment by measuring S_w on a number of tapes of different thickness and figure 5.7 shows some data. The switching coefficient for 4/79/17 molybdenum permalloy, which has $\sigma =$

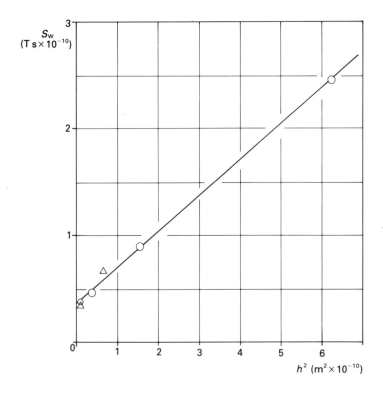

Figure 5.7 Showing how the switching coefficient of 4/79/17 Mo/Ni/Fe tapes varies with the square of the tape thickness. Data points O are from Menyuk (1955) and points △ from O'Dell (1958)

1.86 x 10^6 mho/m and $\mu_0 M_s$ = 0.87 T, is found to vary as h^2 and the slope of the line drawn through the data points in figure 5.7 is 0.33. According to equation 5.33 the slope should be $\sigma\mu_0{}^2 M_s/4$ = 0.51 so that the model is reasonably satisfactory.

The smallest values of S_w shown in figure 5.7 are for tapes 3 μm thick and it is clear that any further reduction in thickness will not produce any improvement in switching speed. This means that the initial part of the switching process, shown in figures 5.6a–5.6b, provides the major part of the magnetic reversal, in these very thin tapes, there being very little of the total volume left, after this initial process, to be reversed by the kind of wall motion shown in figure 5.6b–5.6c. Menyuk (1955) concluded that the kind of results shown in figure 5.7 meant that eddy-current damping was negligible in sheets under 3 μm in thickness and that the damping mechanism below 3 μm was of the Landau–Lifshitz form, that is, the same damping mechanism as for insulators. This conclusion was supported by some FMR results on materials, very similar to permalloy, which had been reviewed by Kittel (1951) but these early FMR results were subject to the same misinterpretation which was discussed in the previous section. It is clear today that the residual value of $S_w \approx 0.4$ x 10^{-10}, shown in figure 5.7, is simply due to the initial part of the switching process, the K of equation 5.33. This does not depend to any great extent upon the tape thickness.

One final aspect of magnetic reversal in thin metallic tapes is of interest and this is apparent when their switching behaviour is compared with that of a square loop ferrite core of the kind which was used in computer core memories. The ferromagnetodynamics of these now obsolete memory devices is very specialised and we shall not go into details here. At low drive fields, a model of domain nucleation and growth is useful (Menyuk and Goodenough, 1955) while at higher drive fields, ferrite cores seem to fit a rotational model of magnetisation reversal (Gyorgy, 1960, 1963) very similar to the one we shall consider in section 5.7 for thin films. A good review of the literature and theory of this problem is found in the book by Rhein (1974) on digital storage techniques. To return to our comparison between ferrite and thin metal tapes; this is made in figure 5.8 where the switching characteristics of two different tapes are shown and these illustrate the linear relationship between the reciprocal reversal time, τ^{-1}, and the drive field, in accordance with equation 5.31. Both characteristics extrapolate back to cut the drive field axis at around 0.06 mT in figure 5.8 but this is well above the coercive field of these materials, which is about 1.0 μT. The ferrite core, in contrast, shows a switching characteristic which cuts the drive field axis in figure 5.8 very close to its true coercive field of 0.15 mT.

It follows that B_0, in equation 5.31, is very close to the coercivity in the case of a ferrite core but very much greater in the case of the metal tapes. This is exactly the same problem which was discussed in the previous section in connection with equations 5.16 and 5.17. It means that there must be a second kind of magnetic reversal process possible in these tape cores for fields which only just exceed the true coercivity, up to the smallest fields shown in figure 5.8. For

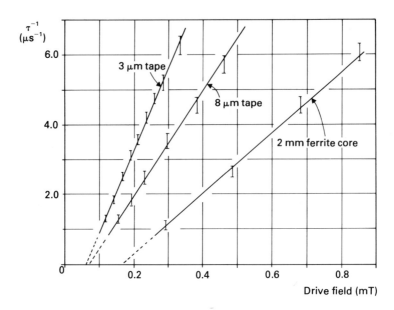

Figure 5.8 The switching characteristics of 3 μm and 8 μm thick 4/79/17 molyb-
          denum permalloy tape are compared with the characteristic for a
          square-loop ferrite core, type number FX1508D2, Mullard Ltd
          (O'Dell, 1958)

the very small fields we have already covered this kind of ferromagnetodynamics
in section 5.2 and the results of Rodbell and Bean (1955), one of which was
included in figure 5.2, show the kind of behaviour involved. In contrast to the
two domain walls approaching one another through the thickness of the tape, as
in figure 5.6c, this slow magnetic reversal takes place by means of two walls,
both of the kind shown in figure 5.1, approaching one another across the width
of the tape.

   The details of magnetic reversal at fields in between the very small fields,
applicable to figure 5.1, and the higher fields, of figure 5.6, are of some theoretical
interest and may be important in understanding the loss mechanism in some
practical cases. Problems of this kind have been reviewed in a paper by Bishop
and Williams (1977) who give many references to related work and describe a
computer simulation of the kind of domain wall motion illustrated here in figure
5.6a. There are some difficulties attached to the detailed solution of these problems, however, which bring us back to section 5.3. The magnetic field produced
by the eddy-currents must be calculated at every point upon the moving domain
wall and then the shape of the wall modified in the light of this calculation. Again,
the eddy-currents and the field they produce must be calculated and convergence
sought. The conditions for such convergence are still a matter of debate (Bishop,
1977, Carr, 1977). The only cases which we have considered here are, firstly,

very low drive in sections 5.2 and 5.4, where the field at the wall is not strong enough to cause any distortion or change in wall shape, secondly, much larger drive fields, in section 5.5, in cases where the sample geometry would make it possible to assume what shape the moving wall might take.

5.6 Domain Wall Inertia in Conducting Media

The suggestion that the eddy-currents associated with domain wall motion may produce additional inertial effects seems to have been made only quite recently by Carr (1976b). It turns out that these inertial effects are only a different way of looking at a rather familiar time-dependent effect; the growth of current in an inductive circuit. Nevertheless, the idea is interesting and can be useful.

Carr (1976b) suggested that the energy associated with the magnetic field produced by the eddy-currents around the moving wall should be added to the total energy of the system. This may be done very simply by using equation 5.5, replacing π by 4 as we did to form equation 5.8, and obtaining an expression

$$(B_e)_{r=0} = \mu_0^2 \sigma M_s v_y h/4 \tag{5.34}$$

for the magnetic field due to the eddy-currents evaluated at the centre of the moving wall shown in figure 5.1b.

The energy density associated with the field B_e is simply $B_e^2/2\mu_0$ and we make the simplifying assumption that this field exists within a cylinder, radius $h/2$, which encloses the moving wall. The total energy which should be associated with a unit length of wall is then

$$E = \pi \mu_0^3 \sigma^2 h^4 M_s^2 v_y^2/128 \tag{5.35}$$

and this may be identified with a kinetic energy, $\frac{1}{2}mv^2$, where m is the mass per unit length of wall. Following Carr (1976b), however, and to correspond to the concept of the Döring (1948) mass, which was introduced in section 2.14, as

$$m_D = 2/\mu_0 \gamma^2 \Delta \tag{5.36}$$

we divide equation 5.35 by h to obtain the energy per unit area of wall and thus obtain

$$m_e = \pi \mu_0^3 \sigma^2 h^3 M_s^2/64 \tag{5.37}$$

as the effective mass per unit area of wall due to the eddy-currents.

Carr (1976b) compared m_e and m_D and found that m_e could exceed m_D in thin sheets once the thickness exceeded about 10 μm. A comparison with the Döring (1948) mass is certainly valid in material as thick as this, when the magnetisation lies in-plane, whereas it is incorrect to use equation 5.36 for very thin films (Aharoni, 1976, 1979). In order to assess the importance of m_e, however, it is more direct to consider the equation of motion

$$m_e \frac{dv}{dt} + \beta v = f(t) \tag{5.38}$$

which states that the sum of the inertial force and the viscous drag upon the wall must always be in balance with any externally applied time varying force $f(t)$. Equation 5.38 tells us that the wall will respond to changes in $f(t)$ with a characteristic time constant

$$\tau = m_e/\beta \tag{5.39}$$

where β is related to the wall mobility by the usual expression

$$\beta = 2M_s/\mu_w \tag{5.40}$$

We now take equation 5.8 as our expression for wall mobility in the case of low drive field eddy-current limited motion and combine equations 5.8, 5.37, 5.39 and 5.40 to give

$$\tau = (\pi/32)\ \mu_0\sigma h^2 \tag{5.41}$$

Equation 5.41 may be compared with the inductance divided by resistance time constant of a circuit which consists of a cylindrical tube, radius r, wall thickness d and length l. The inductance of such a tube is

$$L = \pi\mu_0 r^2/l \tag{5.42}$$

while the resistance is

$$R = 2\pi r/\sigma l d \tag{5.43}$$

The time constant

$$L/R = \mu_0\sigma r d/2 \tag{5.44}$$

gives some insight into our result, equation 5.41 because we have assumed that the eddy-currents occupy the whole volume of a cylinder, radius $h/2$, which surrounds the moving wall. It follows that $d \approx h/2$ and $r \approx h/2$, in equation 5.44, which then gives much the same value to τ as equation 5.41.

In general, an approximate guide to this kind of problem is to argue that the characteristic time constant associated with the eddy-currents is of the order of $\mu_0\sigma$ multiplied by the area enclosed by the eddy-current path. For the moving wall of the kind shown in figure 5.1b this time constant is of the order of 2 ns for the 50 μm Metglas tape, described in table 5.1, but 4 μs for the thicker and higher conductivity SiFe. In both cases, however, these time constants are much shorter than the time scales involved in the experiments (O'Handley, 1975 or Williams *et al.*, 1950).

If we apply the same argument to the kind of wall motion shown in figure 5.6b we would attribute a time constant of the order of $\mu_0 \sigma W h$ to the eddy-currents in this case and the importance of this begins to be apparent when we consider the magnetic reversal of tapes a few millimetres wide and several microns thick. For example, one of the cores discussed in connection with figures 5.7 and 5.8 was 3 μm thick and the tape was 3.5 mm wide. The material 4/79/17 MoNiFe has a conductivity of 1.8×10^6 mho/m. This means that $\mu_0 \sigma W h$ is 23 ns which is getting uncomfortably close to the shortest reversal time of 1/6 μs shown in figure 5.8. The same applies to the 8 μm tape discussed in connection with figure 5.8 because this was 1 mm wide.

We shall have occasion to discuss the $\mu_0 \sigma$ (area enclosed by the eddy-current) idea again in connection with the conducting bubble domain films which are the subject of section 5.8.

5.7 Magnetic Reversal in Thin Permalloy Films

The literature covering magnetic reversal or magnetic switching in thin permalloy films involves several hundred scientific papers over the period 1955 to 1965 and reflects the interest there was at that time in applications for high speed digital data storage. This has been reviewed briefly in section 1.5. Only key references of historical interest will be given in what follows, however, and some review papers will be referred to which came out after the two books that cover the topic (Prutton, 1964 and Soohoo, 1965). Most of the work on magnetic reversal in thin ferromagnetic films was highly specialised and it is clear from the literature that a number of very difficult problems did not begin to be resolved until around 1967, by which time the thin ferromagnetic film was no longer of any practical interest for data storage applications and workers in the field were having to direct their attention elsewhere. This meant that a number of problems and their theoretical solutions were never published in full.

To illustrate how events overtook the ferromagnetic thin film memory it is interesting to look back at the 144K memory described by Stein (1969) which represented some ten years of intensive development and took the possibilities of this technology close to their practical limits. Each bit in this 144K memory occupied an area 0.3 mm x 0.2 mm and the cycle time was 100 ns. Integrated circuit semiconductor memories, which represented a relatively new technology at that time, could already do better than this in both density and speed (Tarui *et al.*, 1969) and clearly had the potential of at least two orders of magnitude improvement in density, a promise which has now been exceeded (Yoshimura *et al.*, 1978).

The applications interest had a profound effect upon the kind of experimental set-up which was used to investigate magnetic reversal in thin films. For example, there was a very fundamental difference between the organisation of the magnetic film memory and the ferrite core store, with which it was often compared. This difference came from the fact that a ferrite memory core exhibits a clear threshold drive field, below which no change in its state can occur, whereas the thin film memory element may exhibit behaviour, often referred to as creep, in which a small irreversible change in the state of magnetisation may take place in even the smallest of drive fields. This means that the thin film storage elements which are used to make up the memory cannot also be used as part of the address logic, as the ferrite cores may be used in the well known coincident current mode of address. The thin film memory must be word organised, meaning that each bit in a word must be entered into a line of storage elements and that that whole line must first be selected in some way.

Selecting the word in a thin film memory involves applying fields which are orthogonal to one another and also involves drive field pulses of a critical length. There are two main kinds of memory: destructive read-out (DRO), in which the information is lost when it is read from the store, and NDRO or non-destructive read-out, where the information is preserved and the need to cycle the data back into the memory is avoided. Fundamental investigations into magnetic reversal in thin films were made using thin film elements in the environments of these two kinds of memory and these are illustrated in figures 5.9 and 5.10.

Figure 5.9 shows the essential environment of a DRO thin film memory element. Magnetic reversal to and from the '1' and '0' states is effected by means of very short write, B_w, and read, B_r, pulses directed along the easy axis of the element. These pulses would be too short to cause any change in the magnetic state if it were not for the constant field, B_0, which is applied perpendicular to the easy direction and causes M to cant away from the direction of B_w and B_r by

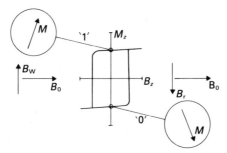

Figure 5.9 Illustrating the two possible states which a small circular thin film element would have in a memory with destructive read-out. The write field B_w, and the read field, B_r, are pulsed. The magnetic switching can only take place in the time allowed by B_w or B_r when the transverse field, B_0, is also applied

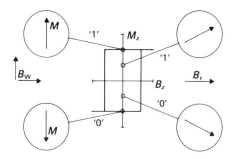

Figure 5.10 In the non-destructive read-out memory the elements are interrogated by applying a pulsed field, B_r, perpendicular to the easy direction

the angle shown in figure 5.9. With this initial condition, the transition '1' $\rightleftarrows$ '0' may be effected by the very fast process of almost coherent rotation of M in the plane of the thin element. The field B_0 is, of course, the word select field in a memory of the DRO kind but for experimental work on a single element a constant field is used.

Figure 5.10 shows the essential environment of an NDRO thin film memory element. Writing is achieved in this store using the same technique described above for the DRO memory and the process of writing '1' is illustrated on the left-hand side of figure 5.10: fields B_w are applied so that the net drive is at an angle to the easy direction and fast coherent rotation is obtained.

Reading from the NDRO memory is quite different compared to the DRO case. As shown in figure 5.10, the read field B_r is applied perpendicular to the easy axis and causes M to rotate well away from the easy direction and either a drop, for the '1', or and increase, for the '0', in the component M_z will be produced. Sensing the read-out signal in this memory is done by inductive coupling with the z-axis so that '1' and '0' read-out signals have opposite signs when gated over the rise time of the read pulse. In contrast, the DRO memory is intended to give zero output when a '0' is read-out.

For both kinds of memory, the fundamental ferromagnetodynamic problem we are interested in is the same: how does the magnetisation of a thin film element change when a pulsed field is applied at an angle to the easy axis. Figure 5.11 shows this problem. Initially, M is at an angle θ_0 to the easy axis, z, because of the field B_0. The drive field, B_z, is then applied and M rotates towards a new equilibrium position, which will be where M makes the same angle, θ_0, with the negative z-axis if B_z is a pulsed field.

Let us first consider the static equilibrium of this problem and the important fact that B_z must exceed some threshold level before any reversal is possible anyway, under this single domain model. The model is one in which the memory element is considered to be circular, as illustrated in figures 5.9 to 5.11, perhaps 1 mm in diameter, and about 0.1 μm thick, so that the single domain hypothesis

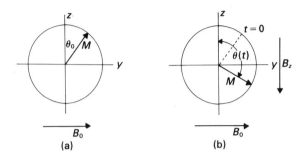

Figure 5.11 Switching a circular thin film memory element. The initial condition (a) is that B_0 has been applied, from $t < 0$, to set M at θ_0 to
the easy axis, z. A pulsed field B_z is then applied and θ increases as
M switches, (b)

of Kittel (1946b, 1949) may be invoked. The effective field vector, F, in the
equation of magnetostatic equilibrium

$$M \times F = 0 \qquad (5.45)$$

is then simply

$$F_x = 0 \qquad (5.46)$$

$$F_y = \mu_0 M_y + B_0 \qquad (5.47)$$

$$F_z = \mu_0 (1 + Q) M_z \qquad (5.48)$$

being the sum of the magnetostatic field, anisotropy field and the field, B_0,
applied in the plane of the element and perpendicular to the easy direction of
magnetisation, z.

Referring to figure 5.11a we have

$$M_y = M_s \sin \theta_0 \qquad (5.49)$$

and

$$M_z = M_s \cos \theta_0 \qquad (5.50)$$

which, substituted into equations 5.45, 5.47 and 5.48 give

$$\theta_0 = \arcsin (B_0/\mu_0 Q M_s) \qquad (5.51)$$

as the angle between M and the easy axis which is brought about by the in-plane
field B_0.

We now turn to figure 5.11b and add the field B_z to equation 5.48 and obtain

$$F_z = \mu_0(1 + Q) M_z + B_z \qquad (5.52)$$

Generalising θ_0 to θ in equations 5.49 and 5.50, these are submitted into equation 5.45 again, with equations 5.47 and 5.52, to obtain a general relationship between θ and the field B_z as

$$B_z = -\mu_0 Q M_s \ (\cos \theta - \sin \theta_0 \ \cot \theta) \qquad (5.53)$$

where we have made use of equation 5.51.

Equation 5.53 is represented in figure 5.12 by plotting not θ against B_z but $M_z/M_s = \cos \theta$ against $B_z/\mu_0 Q M_s$. This gives us what is essentially the quasistatic magnetisation characteristic of the thin film memory element. The field B_0 has been chosen, quite arbitrarily, to make $\theta_0 = 30°$ for figure 5.12.

Magnetic reversal, or switching, of the element must occur when θ, figure 5.12b, exceeds the value θ_c at which $dB_z/d\theta = 0$. Differentiating equation 5.53 gives the relationship between θ_c and θ_0 as

$$\sin^3 \theta_c = \sin \theta_0 \qquad (5.54)$$

and from equations 5.51 and 5.53 we may write

$$\sin^3 \theta_c = B_0/\mu_0 Q M_s \qquad (5.55)$$

and

$$\cos^3 \theta_c = B_{zc}/\mu_0 Q M_s \qquad (5.56)$$

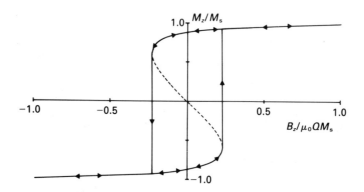

Figure 5.12 The relationship between $\cos \theta$ and $B_z/\mu_0 Q M_s$ given by equation 5.53 results in the S-shaped curve, shown here by a broken line where the slope is negative because, in fact, the single domain must switch at points where the slope is infinite

relationships which were first given by Smith (1958) and by Olson and Pohm (1958), although a number of later authors pointed out that these results had been given in a much more general treatment by Stoner and Wohlfarth (1948). A plot of B_{zc} against B_0, made by using equations 5.55 and 5.56 became known as the switching astroid, or Stoner–Wohlfarth diagram, and showed how the field required to just produce reversal, B_{zc}, would be reduced from the value $\mu_0 Q M_s$ as B_0 was increased from zero. Very good agreement with this simple single domain theory was reported by Smith (1958) over the range $0.1 < B_0/\mu_0 Q M_s < 0.77$.

We now turn to consideration of the problems of dynamics, as opposed to the simple quasistatic situations which have been discussed so far. To begin with, it is interesting to examine just how badly the single domain theory fits the problem. This model assumes that the magnetic reversal takes place by means of a coherent or uniform rotation of M in the plane of the film. Experimental measurements by Humphrey (1958) and by Dietrich *et al.* (1960) had shown that M certainly did have a component which rotated in the plane of the film during reversal of the magnetisation under conditions of high drive but that this component did not have a constant magnitude, M_s. Hearn (1964) was able to observe quite a high degree of coherent rotation under certain switching conditions.

The conductivity of the film can have no influence on the dynamics if uniform or coherent rotation of M is assumed because the time constant which would be associated with the eddy-currents produced, $\tau = \mu_0 \sigma D h$ from the argument given in section 5.6, will be of the order of 10^{-10} s for an element with diameter $D \approx 10^{-3}$ m and thickness $h \approx 10^{-7}$ m. This follows because $\mu_0 \sigma \approx 1$ for these alloys. The magnetisation is moving very slowly on a time scale of 10^{-10} s and we turn to the simple Landau–Lifshitz equation, discussed at length in previous chapters, as the simplest model to use. The form given as equation 2.36 in chapter 2, and attributed to Gilbert and Kelly (1955)

$$\frac{-1}{|\gamma|} \frac{dM}{dt} = M \times \left(F - \frac{\alpha}{|\gamma| M_s} \frac{dM}{dt} \right) \qquad (5.57)$$

is the simplest form to use in this case because if we assume that reversal takes place by means of a rotation of M in (y, z), figure 5.11, $dM_x/dt \approx 0$ and equation 5.57 reduces to

$$\frac{\alpha}{|\gamma|} \frac{d\theta}{dt} + B_z \sin \theta + \mu_0 Q M_s (\sin \theta - \sin \theta_0) \cos \theta = 0 \qquad (5.58)$$

when we substitute equations 5.47, 5.49, 5.50 and 5.52 into equation 5.57 and use equation 5.51.

Equation 5.58 is very easy to solve numerically and solutions were given by Smith (1958), Olson and Pohm (1958) and Matsumoto *et al.* (1966) which, when compared with the experimental facts, immediately showed that something was quite wrong with the coherent rotation model for thin film switching. If we compare equation 5.53 and the last two terms on the left-hand side of equation 5.58 it is clear, particularly when the representation of equation 5.53 given in figure 5.12 is considered, that solutions of equation 5.58 must involve a much larger value of $d\theta/dt$ during the second half of the reversal process, when B_z and $\sin \theta$ both have the same sign, than during the first half.

It follows that we would expect the voltage pulse, induced in a sense coil coupling with the z-axis, to show a slow rise followed by a very rapid fall. Experimentally, just the opposite is observed and figure 13 of Smith (1958) illustrates this point well by comparing the solutions with experiment. Figure 14 of the paper by Matsomoto *et al.* (1966) gives a good picture of the form taken by the numerical solutions.

Before looking at more accurate models of magnetic reversal in thin film memory elements it is interesting to consider a remarkably simple analytical solution to equation 5.58 which can be found when the drive field is assumed to be very high and does correspond with what is observed experimentally provided a suitable value of the damping factor, α, is chosen. When $B_z \gg \mu_0 Q M_s$, it may be argued that equation 5.58 is approximately the same as

$$\frac{d\theta}{dt} = -\frac{|\gamma|B_z}{\alpha} \sin \theta \qquad (5.59)$$

A solution for $\theta(t)$ is not very useful, however, and it would be better to find dM_z/dt because this will be proportional to the voltage pulse, discussed above, which is observed experimentally.

Because $M_z = M_s \cos \theta$, in general, we may write

$$\frac{d(M_z/M_s)}{dt} = -(\sin \theta)\frac{d\theta}{dt} \qquad (5.60)$$

and substitute this into equation 5.59 to obtain

$$\frac{d(M_z/M_s)}{dt} = \frac{|\gamma|B_z}{\alpha} [1-(M_z/M_s)^2] \qquad (5.61)$$

A solution to equation 5.61 may be obtained by substituting

$$M_z/M_s = \tanh x \qquad (5.62)$$

into equation 5.61 to give

$$dx/dt = |\gamma|B_z/\alpha \qquad (5.63)$$

and integrating to obtain

$$x = (|\gamma|B_z/\alpha)t + C \tag{5.64}$$

By choosing $x = 0$ at $t = 0$ the constant of integration is made zero and equation 5.64 put back into equation 5.62 and then equation 5.61 to give

$$\frac{d(M_z/M_s)}{dt} = \frac{|\gamma|B_z}{\alpha} \, \text{sech}^2 \left(\frac{|\gamma|B_z t}{\alpha} \right) \tag{5.65}$$

as an analytic solution for the form of the voltage pulse. Our choice of $x = 0$ at $t = 0$ gives this very simple result, symmetrical about $t = 0$, and implies that the switching process started at $t = -\infty$ and continues to $t = +\infty$. The width of the voltage pulse, taken between its half peak amplitude points, is found by setting $|\gamma|B_z t_s/\alpha = 0.88$, because sech $(0.88) = 1/\sqrt{2}$, where $2t_s$ is the switching time, τ. It follows that

$$\tau = 1.76\alpha/|\gamma|B_z \tag{5.66}$$

a result first given by Conger (1956).

It is certainly true that the voltage pulse observed from thin film memory elements does take on the symmetrical sech^2 (ωt) form at high drive, as shown by the waveforms given by Dietrich *et al.* (1960). These authors observed a pulse width, τ, of 10^{-9} s at a drive field of 0.6 mT and these figures, substituted into equation 5.66 imply a value of $\alpha = 0.06$. This seemed to be a reasonable result at that time but more recent work (Patton *et al.*, 1975) would suggest that a value of $\alpha = 0.005$ would be closer to the truth. The solution, equation 5.65, should be considered of historical interest more than anything else.

A considerable time elapsed before the problem of what was really happening during the magnetic reversal of a thin film element was understood. Experiments of the kind described by Sakuri *et al.* (1966) and by Hoper (1967) showed that the magnetisation of the film could rotate in-plane, and maintain a constant amplitude, over quite a large angle, certainly up to and even over 90°. At some angle, the angle which became known as the locking angle, there would be a sudden drop in the speed of rotation which was usually accompanied by a drop in the in-plane component of *M*.

An explanation for this kind of behaviour began to emerge once it was realised that the initial state of magnetisation within a thin film memory element was not one of simple uniform saturation but one which involved 'magnetisation ripple'. This term had been introduced by Fuller and Hale (1960) to explain the fine texture they had observed in Lorentz electron micrographs of thin permalloy films. Figure 5.13 illustrates what is meant. A thin film memory element is normally found divided into regions within which the magnetisation is

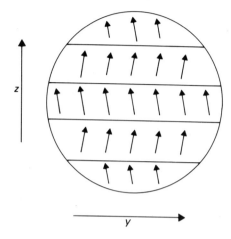

Figure 5.13 **Illustrating magnetisation ripple in a thin film memory element.
The ripple wavelength has been grossly exaggerated relative to the
element diameter, which would perhaps be 1 mm, in order to make
the picture clear**

tilted by a small angle, usually in the plane of the film, away from the macro-
scopic easy direction of magnetisation. The regions alternate, as shown in figure
5.13, with a periodicity of approximately four times the crystallite size and this
showed that the origin of magnetisation ripple must lie in a random component
of *M*, perpendicular to the film easy axis, associated with each crystallite in the
polycrystalline evaporated film (Callen *et al.*, 1965). There is an excellent review
of magnetisation ripple by Leaver (1968), a critical assessment of some aspects
of the micromagnetic theory by Brown (1970) and work in this field continues
(Timakova *et al.*, 1977) because the exact mechanism which produces all the
features of the ripple structure, particularly the longer wavelength components,
is still not completely understood.

The effect which magnetisation ripple has upon the magnetic reversal process
in thin permalloy films, and in thin films with similar magnetic properties, has
been the subject of a very large number of papers. A review of dynamics has
been given by Hagedorn (1968) in an important issue of the *IEEE Transactions
on Magnetics* which gives an account of the Third International Colloquium on
Magnetic Films, Boston, 1967, and includes other important papers, Hoffmann
(1968), Slonczewski (1968) and Weber (1968), with very interesting discussions,
to which reference may have already been made. This meeting at Boston really
marked the beginning of a deeper understanding of thin film ferromagnetodynamics.

The problem may be best summarised by reference to the paper by Hoffmann
and Kryder (1973). Figure 5.14 shows the kind of process which must occur
during magnetic reversal of a thin film memory element which initially has

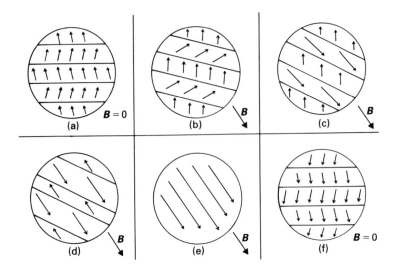

Figure 5.14 **The process of magnetic reversal in a thin film memory element**
         when ripple is included. Again, the ripple wavelength is exaggerated
         for clarity

magnetisation ripple. As in figure 5.13, the ripple wavelength has been exaggerated, relative to the diameter of the element, so that the state of magnetisation and its time dependence may be illustrated more clearly.

Initially, as shown in figure 5.14a, the magnetisation is directed, on average, along the easy direction. When a magnetic field, B in figure 5.14b, is applied in order to reverse the direction of magnetisation, some parts of the film will respond much more rapidly than others because the ripple causes M and B to have a different angle relative to one another in different parts of the film.

As shown in figure 5.14b, the B-field causes M to rotate clockwise in all parts of the film but the rotation is much slower in the centre and the extreme edges of the film because the ripple causes M and B to be very nearly antiparallel in these regions. Where $(M \times B)$ has a greater magnitude, M rotates more rapidly, but the average M of the entire film does rotate, as shown in figure 5.14a to b, while the average amplitude of M falls. Eventually, as in figure 5.14c, the magnetisation within the regions which rotate rapidly will become unstable, due to the magnetostatic field from the divergence in the magnetisation shown in figure 5.14b, and these regions will reverse on their own producing what is now, very nearly, a multidomain thin film element.

Once magnetic domains have been generated, as shown in figure 5.14c, the magnetic reversal of the entire film can proceed by means of domain wall motion, figure 5.14d, until the film becomes saturated, as shown in figure 5.14e. The drive field, B, may then be turned off and the thin film element relaxes back to allow the average M to lie along an easy direction again with the same ripple feature which it had initially. This reversed state is shown in figure 5.14f.

The magnetostatic fields originating from the div M of the distributions shown in figures 5.14b and c play a very significant part in this model and are responsible for the fact that M reverses its sense of rotation in the narrow domains shown in figure 5.14d. Hoffmann and Kryder (1973) were able to verify a considerable amount of detail in the reversal process over a wide range of applied fields and to relate their observations back to the high speed photographs which had been taken by Kryder and Humphrey (1969a, 1969b, 1970), particularly the angle made by the domain walls, shown in figure 5.14d, relative to the film's easy axis. This angle increases with drive in a predictable way. The conclusion of this work by Hoffmann and Kryder (1973) was that this model answered most of the problems which had been brought up by thin ferromagnetic film reversal or switching and brought to a close what had been a very interesting and active line in applied magnetism.

5.8 Wall Motion in Conducting Bubble Domain Films

Amorphous metallic films were proposed for use in magnetic bubble domain devices by Chaudhari *et al.* (1973) and the system which was first tried, and developed later to a limited extent, was based upon gadolinium and cobalt. These two elements form an amorphous ferrimagnet when sputtered under the correct conditions and an easy direction of magnetisation forms normal to the plane of the film, that is, in the direction of build up or growth during the sputtering process. An account of the preparation and properties of such films, together with an extensive bibliography has been given by Gambino *et al.* (1973).

The eddy-current damping which is associated with domain and domain wall motion is quite different in these films when compared to the permalloy films, which were considered in section 5.4. This is because the easy direction of magnetisation is no longer in-plane. The magnetic field which would surround a straight domain wall in a bubble film has already been illustrated here in figure 3.2 and is a field circulating around the wall with a magnitude $\approx \mu_0 M_s$. This magnetic field will be carried along with the moving wall so that there will be an induced current, $\approx \mu_0 \sigma |v| M_s$ per unit area, just in front of the moving wall and just behind it. The relationship between the wall velocity and the applied field is then easily found and was given by Kryder and Hu (1973) in one of the earliest papers on bubble dynamics in these amorphous films.

It is more interesting to consider the induced current around a bubble domain in translation, than around the rather academic straight domain wall, and this problem is illustrated in figure 5.15. In figure 5.15a a bubble domain is shown moving at velocity v_y and with its magnetisation directed into the paper. From the point of view of a stationary observer in the film there will be an electric field $v \times B$ with the signs shown and a current field will be induced in the film as shown in figure 5.15b. To estimate the effect of this current field we may rep-

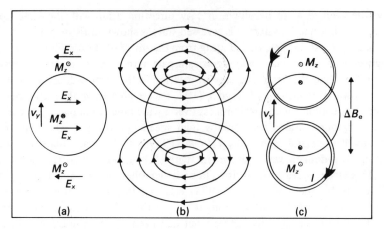

Figure 5.15 **The electric field associated with a bubble domain in translation (a) sets up a pair of circulating current fields, as shown in (b). These may be treated very simply as concentrated current loops, (c)**

resent it by means of two concentrated current loops, as shown in figure 5.15c, and argue that the intensity of these currents must be of the order of $I \approx \mu_0 \sigma v_y hRM_s$, where σ is the conductivity of the bubble film, h is its thickness and R is the bubble radius. The diameter of the current loops is shown to be less than the bubble diameter in figure 5.15c for clarity but if we take the radius of the current loops to be equal to R the magnetic field produced at the centre of these loops will be $\mu_0 I/2R$ and we see at once that the field difference produced across the bubble diameter, ΔB_e in figure 5.15c, will be $\mu_0 I/R$ or, using the expression for I given above

$$|\Delta B_e| = \mu_0{}^2 \sigma v_y hM_s \qquad (5.67)$$

The sign of ΔB_e, in figure 5.15c, is such that it would drive the bubble downwards, that is, it opposes the motion. We now use the same argument used in section 5.2: that ΔB_e must be equal and opposite to the drive, ΔB_z, which would have to be applied to drive the bubble forward at velocity v_y. Referring back to chapter 4, equation 4.42 we have

$$v_y = (\mu_w/2) D (dB_z/dy) \qquad (5.68)$$

when we neglect coercivity, and setting $D (dB_z/dy) = |\Delta B_e|$ in the combination of equations 5.67 with 5.68 an expression for μ_w

$$\mu_w = 2/\mu_0{}^2 \sigma h M_s \qquad (5.69)$$

is obtained which is very similar to the previous expressions for wall mobility, equations 5.8 and 5.24, given for the case of conducting ferromagnetic layers which have M in-plane.

Equation 5.69 predicts a very high mobility, $\mu_w = 1.3 \times 10^7$ m/s per T when typical values for a gadolinium cobalt film, $\sigma \approx 5 \times 10^5$ mho/m, $h \approx 2 \times 10^{-6}$ m and $\mu_0 M_s = 0.12$ T, are substituted (Kryder and Hu, 1973). As Kryder and Hu (1973) pointed out, even the highest mobilities which have been measured in this material are at least a whole order of magnitude smaller than this and later work by Kryder *et al.* (1977) suggested that mobilities between 10^4 and 2.5×10^4 m/s per T were applicable to these amorphous films. The concluison was that eddy-currents, or the effects of conductivity, could be neglected in these bubble domain materials and that the ferromagnetodynamics should be the same as for the garnet materials which were the subject of chapter 4.

Nevertheless, certain aspects of bubble dynamics in conducting amorphous films suggest that things may be different compared to the bubble dynamics for the garnets. For example, Kryder *et al.* (1975) reported the observation of very high domain wall velocities in a chevron expander/detector feature of a 2 μm bubble device, using GdCoMo, when this was viewed using high speed microscopy. The velocity was 300 m/s which seemed to be too high compared to the saturation velocity expected from the theory discussed in section 3.9. Later work by Hafner and Humphrey (1977) confirmed these very high velocities. Unfortunately, interest in these materials declined, because their high temperature sensitivity made them unsuitable for device applications, and these questions remained open.

5.9 The Interaction between Domain Walls and Electric Currents

The idea that the state of magnetisation might be influenced by the flow of an electric current in a magnetic material appears in the experiments of Sixtus and Tonks which were the subject of section 1.3. Tonks and Sixtus (1933a) found that the propagation velocity of a reversed domain, along a twisted wire of NiFe, could be modulated by passing a current down the wire. The velocity would increase for one sign of current but decrease for the other sign. These effects were due to a very complicated interaction between the magnetostriction of the material, coming in through the torsion in the wire, and the helical form taken by M as a result of the θ field, coming from the longitudinal current, combining with the z magnetic field which was normally present in the Sixtus and Tonks kind of experiment. More recently, effects of this kind have been exploited in the 'Twistor' memory array, described by Bobeck (1957) which became a very important method of data storage in the telephone system during the 1960s (McKenzie, 1964).

Straightforward use of the magnetic field which is produced by an electric current flowing in a magnetic material has been applied to make a direct measurement of domain wall energy by Aléonard *et al.* (1963), whose work has been reviewed by Bishop (1979). Another example is the very interesting method of measuring the exchange stiffness constant in a NiFe thin film, by passing a pulsed current through the film, which has been described by Imamura and Chikazumi (1967).

From these straightforward interactions between domains, or states of mag-netisation, and electric currents we turn towards the far more complicated possibilities which may arise when the magnetic material concerned does not have a simple scalar conductivity but a tensor conductivity which itself depends upon the magnetic field within the material. This can be the case when the material shows the Hall effect or is magnetoresistive. Ideas along these lines were first put forward by Berger (1973a, 1973b, 1974) and very similar proposals were made independently by Carr (1974a, 1974b) and his colleagues (Charap, 1974 and Emtage, 1974). The problem has also been considered from a more general point of view by Zaitsev (1976). The conclusion of all these authors was that it should be possible, by selecting the right kind of materials or a combination of such materials, to obtain domain wall motion in the direction of a steady direct current. This effect was termed 'domain drag' by Berger (1973a) and the name was adopted by DeLuca *et al.* (1978) who first observed the effect in thin films of amorphous GdCoMo.

The reason for the existence of a domain drag effect may be seen by reference to figure 5.16. In figure 5.16a the normal Hall effect is shown in that a thin rectangular plate, having a magnetic field, B_z, passing through it, carries a current I. For simplicity we assume that the charge carriers are all positive, with charge e, and consider a single carrier at the centre of the plate, well away from any contact effects. This positive charge carrier has a force eE_y on it, causing it to move up the paper in figure 5.16a with velocity v_y, and has the Lorentz force

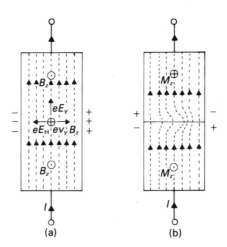

Figure 5.16 **Current flow in a simple Hall plate (a) is compared with that in a ferromagnetic plate, bisected by a single domain wall, (b), with an easy direction of magnetisation normal to the plane of the plate**

$e(v \times B)$, shown as ev_yB_z in figure 5.16a, acting to the right. Because the boundary conditions do not allow a component of current across the plate, the force ev_yB_z must be in equilibrium with the force due to the electric field, E_H, coming from the Hall voltage, represented by the $+$ and $-$ signs on either side of figure 5.16a. It follows that we have

$$E_H = v_yB_z \qquad (5.70)$$

and if we introduce a positive carrier mobility, μ_+, so that

$$v_y = \mu_+E_y \qquad (5.71)$$

equation 5.70 may be written

$$E_H/E_y = \mu_+B_z \qquad (5.72)$$

The dimensionless quantity μ_+B_z is often referred to as the Hall angle (Berger, 1974) because it is the tangent of the angle at which the carriers would move, relative to the applied electric field E_y, if the Hall voltage, and thus E_H, were in some way short-circuited. In practical situations, μ_+B_z, or μ_-B_z if the carriers are electrons, is usually well below unity but this is certainly not the case in very high mobility materials like indium antimonide which may have an electron mobility as high as 50 m/s per V/m at low temperatures.

We now consider figure 5.16b where the Hall plate of figure 5.16a is replaced with a layer of ferromagnetic material having an easy direction of magnetisation normal to the plane of the layer, like a bubble domain film, and we assume that this layer is divided into two domains with a single straight domain wall across the centre of the film and at right angles to the current, I, which is made to flow in the film.

There will be no magnetic field in this layer normal to its plane except around the edges and close to the domain wall where there will be a large B-field, of the order of μ_0M_s, circulating around the wall in (y, z). This field was first shown and discussed here in connection with figure 3.2. This field will try to set up equal and opposite Hall voltages across the film, illustrated in figure 5.16b by the $\pm$ and $\mp$ signs on either side, and at the centre of the film E_H will be effectively reduced to zero. The carriers, again assumed to be positive, will approach the domain wall close to the Hall angle to produce a component of current density, $\mu_+\mu_0M_sJ_y$, on either side of the wall and parallel to the wall. This component of J only exists in a region of order h^2 in cross-sectional area and may thus be represented by the currents $i = \mu_+\mu_0M_sJ_yh^2$ shown in figure 5.17 flowing in opposite directions along either side of the domain wall and separated by a distance $\approx h$. The magnetic field produced at the wall by these two currents will be

$$B_z = 2\mu_0^2\mu_+M_sJ_yh/\pi \qquad (5.73)$$

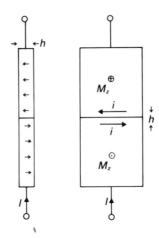

Figure 5.17 (a) A side view of the ferromagnetic plate with its *M*-field. (b) The
same view as figure 5.16b, the components of current ±i produced
along either side of the domain wall by the Hall effect

so that, if the wall has a simple mobility μ_w, the domain wall drag velocity, v_d,
will be

$$v_d = 2\mu_w\mu_+\mu_0{}^2M_sJ_yh/\pi \qquad (5.74)$$

Apart from a numerical factor very close to unity, equation 5.74 is almost
identical to the relationship used by DeLuca *et al.* (1978) to compare their
experimental results with theory. The GdCoMo film used by DeLuca *et al.* (1978)
had $\mu_w = 5 \times 10^4$ m/s per T, $\mu_+\mu_0M_s = 3.6 \times 10^{-3}$ rad and $h = 2.1 \times 10^{-6}$ m. These
values, substituted into equation 5.74, give $v_d/J_y = 1.44 \times 10^{-10}$ m^3/A s, and the
value measured by DeLuca *et al.* (1978) was nearly ten times smaller at 2.5×10^{-11}.
As figures 5.16b and 5.17 show, however, only the central part of the film can
show the domain drag effect and, in practice, a bubble domain array or a set of
stripe domains must be used because the isolated and straight wall is unstable, as
we discussed at length in chapter 3. In view of this the lower value of v_d/J_y
which is observed makes good sense.

The simple theory given above shows that the domain drag velocity must
always be in the direction of the carrier velocity, not in the direction of the conventional current. This could be confirmed in the experiments of DeLuca *et al.*
(1978) because they had several films, some *p*-type and some *n*-type.

The Hall effect is not the only effect to bring about a magnetic field dependent tensor conductivity in a medium, magnetoresistance is another possibility.
It is here that we make contact with the theoretical work of Carr (1974a, 1974b)
Charap (1974) and Emtage (1974) who considered the problem of a magnetic
layer in contact with a semiconducting or magnetoresistive layer. The problem

is very similar to the one we have discussed here but is of far more practical significance because a general principle of good engineering design is being followed when we separate the specification of a system and no longer ask one part to do two different jobs. For example, when the Hall effect layer and the magnetic layer are distinct, we may use indium antimonide for one layer and a good bubble garnet for the other. Useful device applications are immediately obvious in that bubble propagation may be possible by controlling the current flow in the semiconducting or magnetoresistive layer. Such an idea is reminiscent of the interesting device proposed by Walsh and Charap (1974) in which a normal conductor was used as an overlay and the current within this was perturbed, not by using Hall effect or magnetoresistance, but by the simple technique of cutting a specially shaped aperture through the conducting layer and then controlling the direction of the current flowing in the layer. This kind of device technology has now been taken to a high degree of development using as many as three conducting layers over the bubble garnet layer (Bobeck *et al.*, 1979).

5.10 Conclusions

This chapter has been concerned with domain and domain wall motion in conducting ferromagnetic layers, first under very small drive fields and then under much higher drives. The intermediate drive region discussed at the end of section 5.5 is certainly a far more difficult problem and is of great interest mathematically. It is in this region that further developments may be expected and the interesting paper by Becker (1963) recognised this quite early and connects the very early work, referred to at the beginning of section 5.2 and the beginning of section 5.5, with the recent work referred to at the end of section 5.5.

Another area where developments are certainly expected is the subject of the last section, 5.9. This was the motion of domains and domain walls caused by simple direct current flow in the ferromagnet itself or in a composite ferromagnet–conductor. This kind of indirect motional effect is reminiscent of the bubble automotion effects, discussed in chapter 4, section 4.12, and automotion itself is very similar experimentally to another interesting ferromagnetodynamic effect which was first observed in thin permalloy films by Kusuda *et al.* (1967) and by Stein and Feldtkeller (1967). This effect was termed 'wall-streaming' by Feldtkeller and Stein (1968) and involves the motion of a domain wall which occurs when a pulsed magnetic field is applied perpendicular to the wall plane. For a simple Landau–Lifshitz wall, the kind described by equations 2.30 and 2.31, such a pulsed field would be expected to cause a forward motion of the wall, or a backward motion depending upon the wall screw-sense, during the rise time of the field pulse. This forward motion need not necessarily be cancelled out when the pulsed field goes off, either because of coercivity or because the fall time of the drive pulse is made much longer than the rise time. This latter technique was used by Malozemoff and Slonczewski (1974) in their attempt to observe wall-streaming in a bubble garnet film. The effect may be quite spectacular in permalloy thin films and has been made the subject of a motion picture for the *Encyclopedia Cinematographica* (Stein and Feldtkeller, 1967). Wall-streaming was of some importance in the permalloy thin film memory because the word-select field pulse, discussed briefly in section 5.7, was precisely the kind of applied field to cause wall-streaming and thus a degradation of the data stored in some kinds of thin film memory systems.

References and Author Index

The pages in this book where the citation has been made are given within square brackets.

Aharoni, A. (1974), Critique on the theories of domain wall motion in ferromagnets, *Phys. Lett.*, **50A**, 253-4. [180]

Aharoni, A. (1975a, b, c), Two dimensional domain walls in ferromagnetic films. I General theory. II Cubic Anisotropy. III Uniaxial anisotropy, *J. appl. Phys.*, **46**, 908-13, 914-16, 1783-6. [34, 177, 178, 180]

Aharoni, A. (1976), Two dimensional domain walls in ferromagnetic films. IV Wall motion, *J. appl. Phys.*, **47**, 3329-36. [34, 180, 188]

Aharoni, A. (1979), A contribution to the theory of domain wall mass at high velocities, *IEEE Trans. Magnetics*, **15**, 1285-90 [180, 188]

Aharoni, A., and Jakubovics, J. P. (1979), Moving domain walls in ferromagnetic films with parallel anisotropy, *Phil. Mag.*, **40**, 223-31. [180]

Aléonard, R., Brissonneau, P., and Néel, L. (1963), New method to measure directly the 180° Bloch wall energy, *J. appl. Phys.*, **34**, 1321-2. [201]

Argyle, B. E., and Malozemoff, A. P. (1972), Experimental study of domain wall response to sinusoidal and pulsed fields, *AIP Conf. Proc.*, **10**, 344-8. [92]

Argyle, B. E., and Halperin, A. (1973), A measuring system for analyzing AC dynamic domain wall motions in bubble materials, *IEEE Trans. Magnetics*, **9**, 238-42. [15, 92]

Argyle, B. E., Slonczewski, J. C., and Mayadas, A. F. (1971), Domain wall motion in rare-earth substituted Ga:YIG epitaxial films, *AIP Conf. Proc.*, **5**, 175-9. [86, 98]

Argyle, B. E., Slonczewski, J. C., Dekker, P., and Maekawa, S. (1976a), Gradientless propulsion of magnetic bubble domains *J. Magnetism Magn. Mater.*, **2**, 357-60. [23, 151-3, 155]

Argyle, B. E., Slonczewski, J. C., and Voegeli, O. (1976b), Bubble lattice motions due to modulated bias fields, *IBM J. Res. Dev.*, **20**, 109-22. [153, 156]

Argyle, B. E., Maekawa, S., Dekker, P., and Slonczewski, J. C. (1976c), Gradientless propulsion and state switching of bubble domains, *AIP Conf. Proc.*, **34**, 131-7. [24, 124, 151, 153, 155]

Ascher, E. (1966), Magnetic anisotropy: A reformulation and its consequences, *Helv. Phys. Acta*, **39**, 466-76. [30, 67]

Aulock, W. H. von (1965), *Handbook of Microwave Ferrite Materials* (Academic Press, New York and London) [50, 52]

Baldwin, J. A. Jr. (1972), Pressure due to variations in the energy of a magnetic domain wall, *J. appl. Phys.*, **43**, 4830-1. [175]

Baldwin, J. A. Jr., Milstein, F., Wong, R. C., and West, J. L. (1977), Slow magnetic domain wall motion, *J. appl. Phys.*, **48**, 2612-17. [146]

Barbara, B., Magnin, J., and Jouve, H. (1977), Bubble motion at small drive fields, *Appl. Phys. Lett.*, **31**, 133-4. [146]

Bartran, D. S., and Bourne, H. C. (1972), Wall contraction in Bloch wall films, *IEEE Trans. Magnetics*, **8**, 743-6. [64, 173]

Bartran, D. S., and Bourne, H. C. (1973), Domain wall velocity and interrupted pulse experiments, *IEEE Trans. Magnetics*, **9**, 609-13. [64, 173, 180]

Beaulieu, T. J., and Voegeli, O. (1974), Dynamic penetration of potential barriers in garnet films., *AIP Conf. Proc.*, **24**, 627-9. [24]

Beaulieu T. J., and Calhoun, B. A. (1976), Dependence of bubble deflection angle on the orientation of an in-plane magnetic field, *Appl. phys. Lett.*, **28**, 290-2. [137, 145]

Beaulieu, T. J., Brown, B. R., Calhoun, B. A., Hsu, T., and Malozemoff, A. P. (1976), Wall states in ion-implanted garnet films, *AIP Conf. Proc.*, **34**, 138-43. [147, 150, 153]

Becker, J. J. (1963), Magnetisation changes and losses in conducting ferromagnetic materials, *J. appl. Phys.*, **34**, 1327-32. [205]

Becker, R. (1938), *Probleme der Technischen Magnetisierungskurve* (Julius Springer, Berlin) [3, 38]

Becker, R. (1952), Eine Bemerkung zur Massenträgheit der Blochwand, *Z. phys.*, **133**, 134-9. [53-5]

Beljers, H. G. (1949), Measurements on gyromagnetic resonance, *Physica*, **14**, 629-41. [41]

Berger, L. (1973a), Dragging of domains by an electric current in very pure, non-compensated, ferromagnetic metals, *Phys. Lett.*, **46A**, 3-4. [202]

Berger, L. (1973b), Dragging of domains by an electric current in very pure, non-compensated, ferromagnetic metals, *AIP Conf. Proc.*, **18**, 918-22. [202]

Berger, L. (1974), Prediction of a domain drag effect in uniaxial non-compensated, ferromagnetic metals, *Physics Chem. Solids*, **35**, 947-56. [202-3]

Birss, R. R. (1966) *Symmetry and Magnetism* (North Holland, Amsterdam) [68]

Bishop, J. E. L. (1977), Comments on "Magnetic domain wall bowing in a perfect metallic crystal", *J. appl. Phys.*, **48**, 5377-78. [186]

Bishop, J. E. L. (1979), Magnetic domain wall kinking and the Aléonard-Brissonneau-Néel experiment, *J. Phys. D*, **12**, L1-L4. [201]

Bishop, J. E. L., and Lee, E. W. (1963), A domain theory of the skin-effect impedance and complex permeability, *Proc. R. Soc. A*, **276**, 96-111. [165]

Bishop, J. E. L., and Williams, P. (1977), A comparison of rapid surface and volume magnetisation measurements on 50% NiFe tape with models of eddy-current limited domain wall motion, *J. Phys. D.* **10**, 225-41. [186]

Bloch, F. (1932), Zur Theorie des Austauschproblems und der Remanenzerscheinung der Ferromagnetika, *Z. Phys.*, **74**, 295-335. [37]

Bobeck, A. H. (1957), A new storage element suitable for large-sized memory arrays: the twistor, *Bell Syst. tech. J.*, **26**, 1319-40. [201]

Bobeck, A. H. (1967), Properties and device applications of magnetic domains in orthoferrites, *Bell Syst. tech. J.*, **46**, 1901–25. [16, 98, 100]

Bobeck, A. H., and Della Torre, E. (1975), *Magnetic Bubbles* (North-Holland, Amsterdam) [16, 100, 115]

Bobeck, A. H., Danylchuk, I., Remeika, J. P., Van Uitert, L. G., and Walters, E. M. (1970), Dynamic properties of bubble domains, *Proc. Int. Conf. on Ferrites, Kyoto, Japan, July 1970*, 361–4. [19–20]

Bobeck, A. H., Blank, S. L., and Levinstein, H. J. (1972a), Multilayer epitaxial garnet films for magnetic bubble devices – Hard bubble suppression, *Bell Syst. tech. J.*, **51**, 1431–35. [21, 117]

Bobeck, A. H., Blank, S. L., and Levinstein, H. J. (1972b), Multilayer garnet films for hard bubble suppression, *AIP Conf. Proc.*, **10**, 498–502. [117]

Bobeck, A. H., Blank, S. L., Butherus, A. D., Ciak, F. J., and Strauss, W. (1979), Current access magnetic bubble circuits, *Bell Syst. tech. J.*, **58**, 1453–1540. [205]

Bourne, H. C., and Bartran, D. S. (1972), A transient solution for domain wall motion, *IEEE Trans. Magnetics*, **8**, 741–3. [64]

Boxall, B. A. (1974), Orthogonal motion of bubble domains in pulsed field gradients, *IEEE Trans. Magnetics*, **10**, 648–50. [138]

Bozorth, R. M. (1951), *Ferromagnetism* (Van Nostrand, Princeton, N. J.) [167]

Bremmer, H., and Bouwkamp, C. J. (1960), *Balthasar van der Pol. Selected Scientific Papers*, Vols I and II (North Holland, Amsterdam) [2]

Brown, W. F. Jr. (1939), Theory of reversible magnetisation in ferromagnets, *Phys. Rev.*, **55**, 568–78. [3]

Brown, W. F. Jr. (1962), *Magnetostatic Principles in Ferromagnetism* (North Holland, Amsterdam) [28, 120]

Brown, W. F. Jr. (1963), *Micromagnetics* (Wiley Interscience, New York) [28]

Brown, W. F. Jr. (1970), A critical assessment of Hoffmann's linear theory of ripple, *IEEE Trans. Magnetics*, **6**, 121–9. [197]

Brown, W. F. Jr. (1978), Domains, micromagnetics and beyond: Reminiscences and assessments, *J. appl. Phys.*, **49**, 1937–42. [177]

Bulaevskii, L. N., and Ginzburg, V. L. (1964), Temperature dependence of the shape of the domain wall in ferromagnets and ferrolectrics, *Soviet Phys. JETP*, **18**, 530–5. [37–38, 40, 68, 96]

Bulaevskii, L. N., and Ginzburg, V. L. (1970) Structure of domain wall in weak ferromagnets, *J. exp. theor. Phys. Lett.*, **11**, 272–4. [37–8, 68, 96]

Bullock, D. C. (1973), The effect of a constant in-plane magnetic field on magnetic bubble translation in $(YGdTm)_3(FeGa)_5O_{12}$, *AIP Conf. Proc.*, **18**, 232–6. [136, 146]

Bullock, D. C., Carlo, J. T., Mueller, D. W. and Brewer, T. L. (1974), Material and submicron bubble device properties of $(LuSm)_3Fe_5O_{12}$, *AIP Conf. Proc.*, **24**, 647–8. [136, 146]

Cahn, J. W., and Kikuchi, R. (1966), Theory of domain walls in ordered structures III, *Physics. Chem. Solids*, **27**, 1305–17. [68]

Calhoun, B. A., Giess, E. A., and Rosier, L. L. (1971), Dynamic behaviour of domain walls in low-moment yttrium–gallium–iron garnet crystals, *Appl. phys. Lett.*, 18, 287-9. [20]

Callen, H. B. (1958), A ferromagnetic dynamical equation. *Physics Chem. Solids*, 4, 256-70. [42, 68]

Callen, H., and Josephs, R. M. (1971), Dynamics of magnetic bubble domains with an application to wall mobilities, *J. appl. Phys.*, 42, 1977-82. [20, 98, 101]

Callen, H. B., Coren, R. L., and Doyle, W. D. (1965), Magnetisation ripple and arctic foxes, *J. appl. Phys.*, 36, 1064-6. [197]

Callen, H., Josephs, R. M., Seitchik, J. A., and Stein, B. F. (1972), Wall mobility and velocity saturation in bubble-domain materials, *Appl. phys. Lett.*, 21, 366-9. [20, 98]

Cape, J. A. (1972), Dynamics of bubble domains, *J. appl. Phys.*, 43, 3551-9. [98]

Cape, J. A., Hall, W. F. and Lehman, G. W. (1974), Theory of magnetic domain dynamics in uniaxial materials, *J. appl. Phys.*, 45, 3572-81. [Errata: 46, (1975) 2338]. [99]

Carr, W. J. Jr. (1974a), Propagation of magnetic domain walls by a self-induced current distribution, *J. appl. Phys.*, 45, 394-6. [202, 204]

Carr, W. J. Jr. (1974b), Force on a domain wall due to perturbation of current in a magnetoresistive overlay, *J. appl. Phys.*, 45, 3115-16. [202, 204]

Carr, W. J., Jr. (1976a), Magnetic domain wall bowing in a perfect metallic crystal, *J. appl. Phys.*, 47, 4176-81. [168, 172]

Carr, W. J. Jr. (1976b), Magnetic domain wall mass in metallic crystals, *AIP Conf. Proc.*, 34, 108-9. [170, 187]

Carr, W. J. Jr. (1977), Wall bowing and stability, *J. appl. Phys.*, 48, 5379. [186]

Chang, H. (1975), *Magnetic Bubble Technology: Integrated-Circuit Magnetics for Digital Storage and Processing* (IEEE Press, New York) [16, 100]

Chang, H. (1978), *Magnetic-bubble Memory Technology* (Marcel Dekker, New York) [100]

Charap, S. H. (1974), Interaction of a magnetic domain wall with an electric current, *J. appl. Phys.*, 45, 397-402. [202, 204]

Chaudhari, P., Cuomo, J. J., and Gambino, R. J. (1973), Amorphous metallic films for bubble domain applications, *IBM J. Res. Dev.*, 17, 66-8. [199]

Clark, A. E., and Callen, E. (1968), Néel ferrimagnets in large magnetic fields, *J. appl. Phys.*, 39, 5972-82. [31]

Clover, R. B., Cutler, L. S., and Lacey, R. F. (1972), Coercivity measurement in magnetic bubble garnets, *AIP Conf. Proc.*, 10, 388-92. [125]

Cobb, C. H., Jaccarino, V., Butler, M. A., Remeika, J. P., and Yasuoka, H. (1973), Low temperature Cr^{53} NMR in a ferromagnetic $CrBr_3$ single crystal, *Phys. Rev. B*, 8, 307-18. [40]

Conger, R. L. (1955), Magnetisation reversal in thin films, *Phys. Rev.*, 98, 1752-4. [13]

Conger, R. L. (1956), High frequency effects in magnetic films, *Proc. Conf.*

Magnetism and Magn. Mater, Boston, USA, 1956 (AIEE Special Publication T-91) 610-19. [196]

Conger R. L., and Essig, F. C. (1956), Resonance and reversal phenomena in ferromagnetic films, *Phys. Rev.*, **104**, 915-23. [13]

Conger, R. L., and Moore, G. H. (1963), Direct observation of high-speed magnetization reversal in films, *J. appl. Phys.*, **34**, 1213-14. [15]

Copeland, J. A. and Humphrey, F. B. (1963), Flux reversal by Néel wall motion, *J. appl. Phys.*, **34**, 1211-12. [15]

Craik, D. J., and Myers, G. (1975), Bloch lines and hysteresis in uniaxial magnetic crystals, *Phil. Mag.*, **31**, 489-502. [122]

DeBlois, R. W., and Graham, C. D. Jr. (1958), Domain observations on iron whiskers, *J. appl. Phys.*, **29**, 931-9. [104]

Dekker, P., and Slonczewski, J. C. (1976), Switching of magnetic bubble states, *Appl. phys. Lett.*, **29**, 753-6. [153-5]

Della Torre, E., Hegedüs, C. and Kádár, G. (1975), Wall structure of cylindrically symmetric magnetic domains, *AIP Conf. Proc.*, **29**, 89-90. [71-2, 105]

DeLuca, J. C., and Malozemoff, A. P., (1976), The effects of in-plane fields on ballistic overshoot in the gradient propagation of magnetic bubble domains, *AIP Conf. Proc.*, **34**, 151-3. [164]

DeLuca, J. C., Malozemoff, A. P., Su, J. L., and Moore, E. B. (1977a), The anisotropy dependence of bubble dynamics in EuGaYIG films, *J. appl. Phys.*, **48**, 1701-4. [138-9, 145]

DeLuca, J. C., Malozemoff, A. P., and Maekawa, S. (1977b), Elliptical distortion and bias compensation in the gradient propagation of bubble domains, *J. appl. Phys.*, **48**, 4672-7. [163]

DeLuca, J. C., Gambino, R. J., and Malozemoff, A. P. (1978), Observation of domain drag effect in amorphous GdCoMo films, *IEEE Trans. Magnetics*, **14**, 500-2. [202, 204]

Dietrich, W., Proebster, W. E., and Wolf, P. (1960), Nanosecond switching in thin magnetic films, *IBM J. Res. Dev.*, **4**, 189-96. [14, 194, 196]

Dillon, J. F. Jr. (1963), Domains and domain walls, in *Magnetism*, Vol. III, ed. G. T. Rado and H. Suhl, (Academic Press, New York) 413-64. [58]

Dillon, J. F. Jr., and Earl, H. E. (1959), Domain wall motion and ferrimagnetic resonance in a manganese ferrite, *J. appl. Phys.*, **30**, 202-13. [11, 52]

Döring, W. (1948), Über die Trägheit der Wände zwischen Weisschen Bezirken, *Z. Naturf.*, **3a**, 372-9. [53-5, 187]

Döring, W. (1966) Mikromagnetismus, in *Handbuch der Physik*, Vol. XVIII, part 2, ed. S. Flügge, (Springer Verlag, Berlin) 341-437. [28]

Emtage, P. R. (1974), Self-induced drive of magnetic domains, *J. appl. Phys.*, **45**, 3117-22. [202, 204]

Enz, U. (1964), Die Dynamik der Blochschen Wand, *Helv. phys. Acta*, **37**, 245-51. [59-63]

Erber, T., and Fowler, C. M. (1969), *Francis Bitter, Selected Papers and Commentaries* (MIT Press, Cambridge, Mass.) [3]

Feldtkeller, E. (1965), Mikromagnetisch stetige und unstetige Magnetisierungs-konfigurationen, *Z. angew. Phys.*, **19**, 530-6. [105, 148-9]

Feldtkeller, E. (1968), Magnetic domain wall dynamics, *Phys. Stat. Sol.*, **27**, 161-70. [64]

Feldtkeller, E., and Stein, K. U. (1968), Wall-streaming: a gyromagnetically induced kind of wall motion, *J. appl. Phys.*, **39**, 863-4. [205]

Filipov, B. N., and Zhakov, S. V. (1975) A contribution to the theory of the dynamic properties of ferromagnetic monocrystalline plates possessing domain structure, *Physics Metals Metallogr.*, **39**, No. 4, 24-35. [165]

Fontana, R. E., and Bullock, D. C. (1976) Temperature dependence of the dynamic properties of $S \approx 1$ bubbles in YSmLuCaGe iron garnet films, *AIP Conf. Proc.*, **34**, 170-1. [126, 137]

Ford, N. C. (1960), Domain wall velocities in thin nickel iron films, *J. appl. Phys.*, **31s**, 300-1s. [14, 169-170, 174, 180]

Forrester, J. W. (1951), Digital information storage in three dimensions using magnetic cores, *J. appl. Phys.*, **22**, 44. [14, 183]

Fuller, H. W., and Hale, M. E. (1960), Determination of magnetisation distribution in thin films using electron microscopy, *J. appl. Phys.*, **31**, 238-48. [196]

Gál, L. (1975), Analysis of domain wall mobility measurement applying translational motion of bubbles, *Acta. phys. hung.*, **38**, 117-22. [21]

Gál, L., and Humphrey, F. B. (1979), High-frequency propagation and failure of asymmetric half-disk field access magnetic bubble device elements, *IEEE Trans. Magnetics*, **15**, 1113-20. [26]

Gál, L., Zimmer, G. J., and Humphrey, F. B. (1975), Transient magnetic bubble domain configurations during radial wall motion, *Phys. Stat. Sol. (a)* **30**, 561-9. [25, 119]

Gallagher, T. J., and Humphrey, F. B. (1977), Bubble-collapse and stripe-chop mechanism in magnetic bubble garnet materials, *Appl. Phys. Lett.*, **31**, 235-7. [26]

Gallagher, T. J., Ju, K., and Humphrey, F. B. (1979), State indentification and stability of magnetic bubbles with unit winding number, *J. appl. Phys.*, **50**, 997-1001. [26, 156]

Galt, J. K. (1952), Motion of a ferromagnetic domain wall in Fe_3O_4, *Phys. Rev.*, **85**, 664-9. [8, 41, 51]

Galt, J. K. (1954), Motion of individual domain walls in a nickel-iron ferrite, *Bell Syst. tech. J.*, **33**, 1023-54. [8-10, 52, 65, 67]

Galt, J. K., Mathias, B. T., and Remeika, J. P. (1950), Properties of single crystals of nickel ferrite, *Phys. Rev.* **79**, 391-2. [9]

Gambino, R. J., Chaudhari, P., and Cuomo, J. J. (1973), Amorphous magnetic materials, *AIP Conf. Proc.*, **18**, 578-92. [199]

Gilbert, T. L., and Kelly, J. M. (1955), Anomalous rotational damping in ferromagnetic sheets, *Proc. Conf. on Magnetism and Magnetic Materials, Pittsburgh, USA* (AIEE Special Publication T-78) 253-63. [42, 129, 194]

Gossard, A. C., Jaccarino, V., and Remeika, J. P. (1962), NMR in domains and walls in ferromagnetic $CrBr_3$. *J. appl. Phys.*, **33**, 1187-8. [40]

Green, A., and Prutton, M. (1962), Magneto-optic detection of ferromagnetic domains using vertical illumination, *J. scient. Instrum.*, **39**, 244-5. [15]

Griffiths, J. H. E. (1946), Anomalous high-frequency resistance of ferromagnetic metals, *Nature*, **158**, 670-1. [41]

Grzhegorzhevskii, O. A., and Pisarev, R. V. (1974), Magneto-optical investigation of the magnetisation compensation point in gadolinium iron garnets, *Soviet Phys. JETP*, **38**, 312-6. [31]

Gurevich, V. A. (1977), Dynamics of a twisted domain wall in a ferromagnet, *Soviet Phys. Solid St.*, **19**, 1701-6. [157]

Gyorgy, E. M. (1960), Flux reversal in soft ferromagnets, *J. appl. Phys.*, **31s**, 110-7s. [185]

Gyorgy, E. M. (1963), Magnetisation reversal in nonmetallic ferromagnets, in *Magnetism*, Vol. III. ed. G. T. Rado and H. Suhl, (Academic Press, New York) 525-52. [185]

Gyorgy, E. M., and Hagedorn, F. B. (1968), Analysis of domain-wall motion in canted antiferromagnets, *J. appl. Phys.*, **39**, 88-90. [96]

Hafner, D., and Humphrey, F. B. (1977), Magnetic bubble and stripe domain wall velocity in amorphous GdCoMo films, *Appl. Phys. Lett.*, **30**, 303-5. [201]

Hagedorn, F. B. (1968), Review of thin-film switching, *IEEE Trans. Magnetics*, **4**, 41-4. [197]

Hagedorn, F. B. (1970), Instability of an isolated straight magnetic domain wall, *J. appl. Phys.*, **41**, 1161-2. [72]

Hagedorn, F. B. (1971), Domain wall motion in bubble domain materials, *AIP Conf. Proc.*, **5**, 72-90. [86]

Hagedorn, F. B. (1974), Dynamic conversion during magnetic bubble domain wall motion, *J. appl. Phys.*, **45**, 3129-40. [89, 99, 106, 160, 163]

Hagedorn, F. B., and Gyorgy, E. M. (1961), Domain wall mobility in single crystal yttrium iron garnet, *J. appl. Phys.*, **32s**, 282-3s. [11]

Haisma, J., van Mierloo, K. L. L., Druyvesteyn, W. F., and Enz, U. (1975), Static conversion of Bloch-line number and the self collapse of hard bubbles, *Appl. Phys. Lett.*, **27**, 459-62. [106]

Hannon, D. M. (1978), Measurement of stripe head velocity using magneto-resistance signal, *J. appl. Phys.*, **49**, 1847-9. [163]

Hansen, P. (1974), Anisotropy and magnetostriction of gallium-substituted yttrium iron garnet, *J. appl. Phys.*, **45**, 3638-42. [85, 127]

Harper, H., and Teale, R. W. (1967), Damping of magnetic domain wall motion by manganic ions in manganese ferrite, *Solid St. Commun.*, **5**, 327-30. [42]

Harper, H., and Teale, R. W. (1969), Damping of magnetic domain-wall motion in pure and ytterbium-doped yttrium iron garnet, *J. Phys. C, Series 2*, **2**, 1926-33. [11, 42, 52, 65, 67]

Harrison, C. G., and Leaver, K. D. (1973), The analysis of two-dimensional domain wall structures by Lorentz microscopy, *Phys. Stat. Sol. (a)*, **15**, 415-29. [39]

Hasegawa, R. (1974), Effect of Bloch points on the dynamic properties of bubble domains, *AIP Conf. Proc.*, **24**, 615-6. [149-50]

Hayashi, N., and Abe, K. (1975), Computer simulation of bubble motion, *Jap. J. appl. Phys.,* **14**, 1705-16. [99]

Hayashi, N., and Abe, K. (1976), Computer simulation of magnetic bubble domain wall motion, *Jap. J. appl. Phys.,* **15**, 1683-94. [99]

Hearn, B. R. (1964), The dynamic behaviour of magnetic thin films, *Int. J. Electron.,* **16**, 33-57. [175, 194]

Hellmiss, G., and Storm, L. (1974), Movement of an individual Bloch wall in a single-crystal picture frame of silicon iron at very low velocities, *IEEE Trans. Magnetics,* **10**, 36-8. [165, 169-70]

Henry, G. R. (1971), Effects of wall inertia on cylindrical magnetic domains, *J. appl. Phys.,* **42**, 3150-3. [98]

Henry, R. D., Besser, P. J., Warren, R. G, and Whitcomb, E. C. (1973), New approaches to hard bubble suppression, *IEEE Trans. Magnetics,* **9**, 514-7. [118]

Henry, R. D., Elliott, M. T., and Whitcomb, E. C. (1976), Bubble dynamics in multilayer garnet films, *J. appl. Phys.,* **47**, 3702-8. [163]

Hoffmann, H. (1968), Theory of magnetisation ripple, *IEEE Trans. Magnetics,* **4**, 32-8. [197]

Hoffmann, H. (1973), Static wall coercive force in a ferromagnetic thin film, *IEEE Trans. Magnetics,* **9**, 17-21. [173, 175]

Hoffmann, H., and Kryder, M. H. (1973), Blocking and locking during rotational magnetisation reversal in ferromagnetic thin films, *IEEE Trans. Magnetics,* **9**, 544-8. [197, 199]

Hoper, J. H. (1967), A study of noncoherent rotational switching for thin magnetic films, *IEEE Trans. Magnetics,* **3**, 166-70. [196]

Hoshikawa, K., Hattandu, T., and Nakamishi, H. (1974), A new method for suppression of hard bubbles in LPE garnet films, *Jap. J. appl. Phys.,* **13**, 2071-2. [118]

Hoshikawa, K., Minegishi, K., Nakamishi, H., Inoue, N., Takimoto, K., and Suemune, Y. (1976), Hard-bubble free garnet films in $(YEu)_3 (FeGa)_5 O_{12}$, *Jap. J. appl. Phys.,* **15**, 387-8. [118]

Hsu, T. (1974), Control of domain wall states for bubble lattice devices, *AIP Conf. Proc.,* **24**, 624-6. [146-147]

Hsu, T., Brown, B. R., and Montgomery, M. D. (1975), The speed difference between an $S = 0$ and an $S = +1$ bubble in the presence of an in-plane magnetic field, *AIP Conf. Proc.,* **29**, 67-9. [147]

Hu, H. L., and Giess, E. A., (1974), Dynamics of micron sized bubbles, *AIP Conf. Proc.,* **24**, 605-7. [136]

Hu, H. L., Beaulieu, T. J., Chapman, D. W., Franich, D. M., Henry, G. R., Rosier, L. L., and Shew, L. F. (1978), 1-K bit bubble lattice storage device: Initial tests, *J. appl. Phys.,* **49**. 1913-17. [146]

Huang, H. L. (1969), Theory of domain-wall mobility in ferromagnetic insulators, *J. appl. Phys.,* **40**, 855-65. [65]

Hubert, A. (1969), Stray-field-free magnetisation configurations, *Phys. Stat. Sol.,* **32**, 519-34. [177]

Hubert, A. (1973), Interactions between Bloch lines, *AIP Conf. Proc.*, **18**, 178–82. [120–121]

Hubert, A. (1974), *Theorie der Domänenwände in geordneten Medien* (Springer Verlag, Berlin) [177]

Hubert, A. (1975), The statics and dynamics of domain walls in bubble materials, *J. appl. Phys.*, **46**, 2276–87. [89, 99, 105, 159]

Hubert, A. (1976), Mikromagnetisch singuläre Punkte in bubbles, *J. Magnetism Magn. Mater.*, **2**, 25–31. [149]

Humphrey, F. B. (1958), Transverse flux change in soft ferromagnets, *J. appl. Phys.*, **29**, 284–5. [14, 194]

Humphrey, F. B. (1975), Transient bubble domain configuration in garnet materials observed using high speed photography, *IEEE Trans. Magnetics*, **11**, 1679–84. [16, 20, 25, 119]

Ikuta, T., and Shimizu, R. (1973), Stroboscopic observation of magnetic domain wall motion with a light emitting diode, *Rev. Scient. Instrum*, **44**, 1412–13. [16]

Imamura, N., and Chikazumi, S. (1967), Determination of exchange stiffness constant of permalloy thin films by a pulse switching technique, *J. phys. Soc. Japan*, **24**, 648. [202]

Iwata, S., Shiomi, S., Uchiyama, S. and Fujii, T. (1979), Complementary σ-bubbles in bubble automotion, *J. appl. Phys.*, **50**, 2195–7. [155]

Joenk, R. J. (1964), Magnetisation reversal in weak ferromagnets, *J. appl. Phys.*, **35**, 919–20. [96]

Josephs, R. M. (1974), Velocity scatter in magnetic bubble motion, *Appl. Phys. Lett.*, **25**, 244–6. [137]

Josephs, R. M., and Stein, B. F. (1974), Velocity scatter in bubble domain transport measurements, *AIP Conf. Proc.*, **24**, 598–600. [137]

Josephs, R. M., Stein, B. F., and Bekebrede, W. R. (1975), Bubble domain translational motion in the presence of an in-plane field, *AIP Conf. Proc.*, **29**, 65–6. [150]

Ju, K., and Humphrey, F. B. (1977a), Transient shapes of magnetic bubble domains during gradient propagation and overshoot, *IEEE Trans. Magnetics*, **13**, 1190–2. [25]

Ju, K., and Humphrey, F. B. (1977b), Gradient propagation, overshoot and creep in magnetic bubble garnet material, *J. appl. Phys.*, **48**, 4656–65. [25]

Ju, K. and Humphrey, F. B. (1979), Gradientless propulsion for S=1 bubble in ion-implanted bubble garnets, *J. appl. Phys.*, **49**, 2212. [156]

Ju, K., Zimmer, G. J., and Humphrey, F. B. (1976), Hard wall sections in gradient propagated magnetic bubble domains, *Appl. Phys. Lett.*, **28**, 741–3. [25]

Kennard, E. H. (1939), Shape of domains in ferromagnets, *Phys. Rev.*, **55**, 312–4. [3]

Kikuchi, R. (1956), On the minimum of magnetisation reversal time, *J. appl. Phys.*, **27**, 1352–7. [13]

Kikuchi, R. (1960), Statistical dynamics of boundary motion, *Ann. Phys.*, **11**, 328–37. [68]

Kinsner, W., Della Torre, E., and Hutton, R. (1974), Bubble cutting circuits, *IEEE Trans. Magnetics,* **10**, 1071-9. [115]

Kittel, C. (1946a), Theory of the dispersion of magnetic permeability in ferromagnetic materials at microwave frequencies, *Phys. Rev.,* **70**, 281-90. [165]

Kittel, C. (1946b), Theory of the structure of ferromagnetic domains in films and small particles, *Phys. Rev.,* **70**, 965-71. [13, 192]

Kittel, C. (1947), Interpretation of the anomalous Larmor frequencies in the FMR experiment, *Phys. Rev.,* **71**, 270-1. [41]

Kittel, C. (1948), On the theory of FMR absorbtion, *Phys. Rev.,* **73**, 155-61. [41]

Kittel, C. (1949), Physical Theory of Ferromagnetic Domains, *Rev. mod. Phys.,* **21**, 541-83. [30, 192]

Kittel, C. (1950), Note on the inertia and damping constant of ferromagnetic domain boundaries, *Phys. Rev.,* **80**, 918. [41]

Kittel, C., (1951), Ferromagnetic resonance, *J. Phys. Radium,* **12**, 291-302. [185]

Kleparskii, V. G., Ilyashenko, E. I., Randoshkin, V. V, and Telesnin, R. V. (1977), Bubble conversion to anomalous behaviour, *Phys. Stat. Sol. (a),* **44**, 771-5. [26, 119, 138]

Kolotov, O. S. and Lobachev, M. I. (1975), Relaxation of the dynamic domain walls arising during pulsating remagnetisation of thin magnetic films in weak fields, *Physics Metals Metallogr,* **40**, No. 2, 181-3. [174, 180]

Kolotov, O. S., Lobachev, M. I., Pogozhev, V. A., and Telesin, R. V. (1975), Motion of dynamic domain walls, *Soviet Phys. JETP Lett.,* **21**, 86-7. [173-4, 180]

Kolotov, O. S., Lobachev, M. I., Musayev, T. S. H., Pogozhev, V. A., and Telesin, R. V. (1978), Concerning the mechanisms of pulsating remagnetisations of thin magnetic films in weak fields, *Physics Metals and Metallogr.,* **43**, No. 5, 85-9. [180]

Kondorsky, E. (1938), On the magnetic anisotropy in ferromagnetic crystals in weak fields, *Phys. Rev.,* **53**, 319 and 1022. [3]

Konishi, S., Hara, S., Sugatani, S., and Sakurai, Y. (1970), Domains and domain-wall motion in grain-oriented 50-percent Ni-Fe tapes, *IEEE Trans. Magnetics,* **6**, 105-10. [170]

Konishi, S., Shoichi, Y., and Kusuda, T. (1971), Domain-wall velocity, mobility, and mean-free-path in permalloy films, *IEEE Trans. Magnetics,* **7**, 722-4. [173, 180]

Krinchik, G. S., and Verkhozin, A. V. (1967), Investigation of the structure of magnetic substances by a magneto-optical apparatus with micron resolution, *Soviet Phys. JETP,* **24**, 890-4. [39]

Kronmüller, H. (1971), Micromagnetic equations for the magnetic charges and the curl of the spontaneous magnetisation, *Z. angew. Phys.,* **32**, 49-53. [28]

Kryder, M. H., and Humphrey, F. B. (1969a), Dynamic Kerr observation of high-speed flux reversal and relaxation processes in permalloy thin films, *J. appl. Phys.,* **40**, 2469-74. [16, 199]

Kryder, M. H., and Humphrey, F. B. (1969b), A nanosecond Kerr magneto-optic camera, *Rev. Scient. Instrum.,* **40**, 829-40. [16, 199]

Kryder, M. H., and Humphrey, F. B. (1970), Mechanisms of reversal with bias fields, deduced from dynamic magnetisation configuration photographs, *J. appl. Phys.*, **41**, 1130-8. [16, 199]

Kryder, M. H., and Hu, H. L. (1973), Bubble dynamics in amorphous magnetic materials, *AIP Conf. Proc.*, **18**, 213-16. [199, 201]

Kryder, M. H., Ahn, K. Y., and Powers, J. V. (1975), Amorphous film magnetic bubble domain devices, *IEEE Trans. Magnetics*, **11**, 1145-7. [201]

Kryder, M. H., Tao, L–J., and Wilts, C. H. (1977), Dynamics of amorphous film bubble devices, *IEEE Trans. Magnetics*, **13**, 1626-31. [201]

Kusuda, T., Konishi, S., and Sakurai, Y. (1967), Worm motion of domain walls in permalloy films, *IEEE Trans. Magnetics*, **3**, 286-90. [205]

LaBonte, A. E. (1969), Two-dimensional Bloch-type domain walls in ferromagnetic films, *J. appl. Phys.*, **40**, 2450-8. [177]

Landau, L., and Lifshitz, E. (1935), On the theory of the dispersion of magnetic permeability in ferromagnetic bodies, *Phys. Z. SowjUn.*, **8**, 153-69. This paper has been reprinted and reset in *L. D. Landau. Collected Papers*, ed. D. ter Haar (Pergamon, Oxford), 101-14, and in *Men of Physics, L. D. Landau*, Vol. 1, ed. D. ter Haar (Pergamon, Oxford) 178-94. [3, 7-8, 17, 27, 34, 40-2, 51, 53]

Leaver, K. D. (1968), Magnetisation ripple in thin ferromagnetic films, *Thin Solid Films*, **2**, 149-72. [197]

Leaver, K. D., and Vojdani, S. (1970), Domain wall motion in single-crystal nickel platelets, *J. Phys. D*, **3**, 729-35. [52]

LeCraw, R. C., Spencer, E. G., and Porter, C. S. (1958), FMR line width in YIG single crystals, *Phys. Rev.*, **110**, 1311-13. [51]

LeCraw, R. C., Blank, S. L., and Vella-Coleiro, G. P. (1975), New high-speed bubble garnets based on large gyromagnetic ratios (high g), *Appl. Phys. Lett.*, **26**, 402-4. [41, 162]

Lee, E. W. (1958), Eddy-current losses in thin ferromagnetic sheets, *Proc. IEE*, **105c**, 337-42. [165]

Lee, E. W. (1960), Eddy-current effects in rectangular ferromagnetic rods, *Proc. IEE*, **107c**, 257-64. [165]

Lee, E. W., and Callaby, D. R. (1958), Direct measurement of the velocity of propagation of a ferromagnetic domain boundary in 'Perminvar', *Nature*, **182**, 254-5. [14, 169-70]

de Leeuw, F. H. (1973), Influence of an in-plane magnetic field on the domain-wall velocity in Ga: YIG films, *IEEE Trans. Magnetics*, **9**, 614-16. [75, 81-2, 84, 87-8]

de Leeuw, F. H. (1974), Domain wall motion in Ga: YIG films, *J. appl. Phys.*, **45**, 3106-8. [82, 138]

de Leeuw, F. H. (1977a), Recent developments in the dynamics of magnetic domain walls and bubbles, *Physica*, **86-88**, B+C, 1320-6. [96]

de Leeuw, F. H. (1977b), Wall velocity in garnet films at high drive fields, *IEEE Trans. Magnetics*, **13**, 1172-4. [82, 96]

de Leeuw, F. H. (1978), An empirical relation for the saturation velocity in bubble domain garnet materials, *IEEE Trans. Magnetics,* **14**, 596–8. [82, 88]

de Leeuw, F. H., and Robertson, J. M. (1974), Bubble velocities in $(YLa)_3$ $(FeGa)_5O_{12}$ films, *AIP Conf. Proc.,* **24**, 601–2. [136]

de Leeuw, F. H., and Robertson, J. M. (1975), Observation and analysis of magnetic domain wall oscillations in Ga: YIG films, *J. appl. Phys.,* **46**, 3182–8. [82-3, 91]

de Leeuw, F. H., van den Doel, R., and Robertson, J. M. (1978), The dynamical behaviour of magnetic domain walls and magnetic bubbles in single-, double-, and triple-layer garnet films, *J. appl. Phys.,* **49**, 768–83. [82-3, 91]

Lermer, R., and Steyerl, A. (1976), Investigation of ferromagnetic domains and Bloch walls by very low energy neutron transmission, *Phys. Stat. Sol. (a)* **33**, 531–41. [39]

Lilley, B. A. (1950), Energies and widths of domain boundaries in ferromagnets, *Phil. Mag.,* **41**, 792–813. [67]

Lin, Y. S., and Keefe, G. E. (1973), Suppression of hard bubbles by a thin Permalloy layer, *Appl. Phys. Lett.,* **22**, 603–4. [118]

Lyle, T. R., and Baldwin, J. M. (1906), Experiments on the propagation of longitudinal waves of magnetic flux along iron wires and rods, *Phil. Mag.,* **12**, 434–68. [2]

Maekawa, S., and Dekker, P. (1976), The 'turn around effect' in bubble propagation, *AIP Conf. Proc.,* **34**, 148–50. [137, 141, 156]

Malozemoff, A. P. (1972), Interacting Bloch lines: a new mechanism for wall energy in bubble domain materials, *Appl. Phys. Lett.,* **21**, 149–50. [21, 104, 122]

Malozemoff, A. P. (1973a), Nanosecond camera for garnet bubble domain dynamics, *IBM tech. Disclos. Bull.,* **15**, 2756–7. [24]

Malozemoff, A.P. (1973b), Mobility of bubbles with small numbers of Bloch-lines, *J. appl. Phys.,* **44**, 5080–9. [22, 136-7]

Malozemoff, A.P. (1976), Theory of saturation velocity and ballistic overshoot for interpreting domain wall oscillation and dynamic bubble collapse experiments in high mobility bubble films, *J. Magnetism Magn. Mater.,* **3**, 234–47. [145]

Malozemoff, A.P., and Slonczewski, J.C. (1972), Effect of Bloch lines on magnetic domain wall mobility, *Phys. Rev. Lett.,* **29**, 952–5. [99, 105, 111]

Malozemoff, A.P., and Slonczewski, J.C. (1974), In-plane pulse response of bubble domains, *AIP Conf. Proc.,* **18**, 603–4. [205]

Malozemoff, A.P., and DeLuca, J.C. (1975), Ballistic overshoot in the gradient propagation of bubbles in garnet films, *Appl. Phys. Lett.,* **26**, 719–21. [24, 137]

Malozemoff, A.P., and Papworth, K.R. (1975), High-speed photography of topological switching and anisotropic saturation velocities in a misoriented garnet film, *J. Phys. D.,* **8**, 1149–55. [96]

Malozemoff, A.P., and Slonczewski, J.C. (1975), Model for transient behaviour in field gradient propagation of bubbles, *IEEE Trans. Magnetics,* **11**, 1091–3. [24, 163]

Malozemoff, A.P., and DeLuca, J.C. (1978), Elliptical distortion mechanism for ballistic overshoot of bubbles, *J. appl. Phys.*, **49**, 1844-6. [163]

Malozemoff, A.P., Slonczewski, J.C. and DeLuca, J.C. (1975), Translational velocities and ballistic overshoot of bubbles in garnet films, *AIP Conf. Proc.*, **29**, 58-64. [137]

Mathias, J.S., and Fedde, G.A. (1969), Plated-wire technology: A critical review, *IEEE Trans. Magnetics*, **5**, 728-51. [14]

Matsumoto, G., Satoh, T., and Iida, S. (1966), Dynamic properties of Permalloy thin films, *J. Phys. Soc. Japan*, **21**, 231-7. [194-5]

McKenzie, A.A. (1964), New era in telephony: electronic switching, *Electronics*, **37**, Oct. 19, 71-86. [201]

Menyuk, N. (1955), Magnetic materials for digital-computer components. II Magnetic characteristics of ultra-thin molybdenum — permalloy cores, *J. appl. Phys.*, **26**, 692-7. [182-5]

Menyuk, N., and Goodenough, J.B. (1955), Magnetic materials for digital-computer components. I A theory of flux reversal in polycrystaline ferromagnetics, *J. appl. Phys.*, **26**, 8-18. [182, 185]

Methfessel, S., Middelhoek, S., and Thomas, H., (1960), Domain walls on thin NiFe films, *IBM J. Res. Dev.*, **4**, 96-106. [104]

Middelhoek, S. (1966), Domain wall velocities in thin magnetic films, *IBM J. Res. Dev.*, **10**, 351-4. [169-70, 173, 177, 180]

Moody, J.W., Shaw, R.W., Sandfort, R.M., and Stermer, R.L. (1973), Properties of $Gd_y Y_{3-y} Fe_{5-x} Ga_x O_{12}$ films grown by LPE, *IEEE Trans. Magnetics*, **9**, 377-81. [92]

Moon, R.M., and Koehler, W.C. (1977), Magnetic form factor of $Sm_{0.76}Y_{0.74}S$ and SmS, *Bull. Am. phys. Soc.*, **22**, 292. [31]

Morris, T.M., and Malozemoff, A.P. (1973), Observation of straight and wavelike domain wall motion in bubble films by high speed photography, *AIP Conf. Proc.*, **18**, 242-6. [24]

Morris, T.M., Zimmer, G.J., and Humphrey, F.B. (1976), Dynamics of hard walls in bubble garnet stripe domains, *J. appl. Phys.*, **47**, 721-6. [121]

Mulhall, B.E. (1964), Eddy current losses in thin magnetic sheet, *Proc. IEE*, **111**, 183-7 and 188-92. [165]

Nakanishi, H., and Uemura, C. (1974), A new type of bubble motion in magnetic garnet films, *Jap. J. appl. Phys.*, **13**, 191-2. [137]

Nedlin, G.M., and Shapiro, R. Kh. (1975), Motion of domain walls in magnetic films, *Soviet Phys. solid St.*, **17**, 1357-62. [99, 159]

Nedlin, G.M., and Shapiro, R. Kh. (1977), Effect of transverse magnetic field on domain-wall motion in ferromagnets, *Soviet Phys. Solid St.*, **19**, 1707-12. [99, 157]

Néel, L. (1951), Influence de la subdivision en domaines élémentaires sur la perméabilité haute fréquence des corps ferromagnétiques conducteurs, *Annls. Inst. Fourier*, **3**, 301-19. [165]

Néel, L. (1955), Energie des parois de Bloch dans les couches minces, *C. R. hebd. Séanc. Acad. Sci., Paris*, **241**, 533-6. [38]

Neilson, J.W., and Dearborn, E.F. (1958), The growth of single crystals of magnetic garnets, *Physics Chem. Solids*, 5, 202-7. [11]

Nishida, H., Kobayashi, T., and Sugita, Y. (1973), Formation of normal and hard bubbles by cutting strip domains, *IEEE Trans. Magnetics*, 9, 517-20. [115-17]

North, J.C., Wolfe, R.C., and Nelson, T.J. (1978), Applications of ion-implantation to magnetic bubble devices, *J. vac. Sci. Technol.*, 15, 1675-84. [117-18, 133]

Oberbeck, A. (1884), Ueber electrische Schwingungen. Die magnetisirende Wirkung derselben, *Annln Phys. Chem.*, 21, 672-97. [2]

Obokata, T., Yamaguchi, K., and Asama, K. (1975), Temperature stability of bubble domain wall states, *AIP Conf. Proc.*, 29, 74-5. [137, 147]

O'Dell, T.H. (1958), Molybdenum nickel iron as a material for magnetic storage applications, *UKAEA Research Report, AERE-EL/R-2595*. [184]

O'Dell, T.H. (1970), *The Electrodynamics of Magneto-electric Media* (North Holland, Amsterdam) [68]

O'Dell, T.H. (1973), The dynamics of magnetic bubble domain arrays, *Phil. Mag.*, 27, 595-606. [15]

O'Dell, T.H. (1974), *Magnetic Bubbles* (Macmillan Press Ltd, London; Halstead Press, New York, 1975; in Russian, *Magnite Domeny vysokoi podvizhnosti*, Izd. Mir, Moscow, 1978.) [16, 20, 69, 100-2, 125, 143]

O'Dell, T.H. (1975), A review of some dynamic experiments with magnetic bubble domains, *Proc. Winter School on New Magnetic Materials, Kocierz, Poland, April 20-30, 1975*, 101-12. [136]

O'Dell, T.H. (1978a), Static wall structure in bubble films under large in-plane fields, *Phys. Stat. Sol. (a)*, 48, 59-66. [77, 92, 157]

O'Dell, T.H. (1978b), The peak velocity of a domain wall in a magnetic bubble film under a large in-plane field, *Phys. Stat. Sol. (a)*, 50, 179-86. [85]

O'Dell, T.H. (1979), Domain wall mass in magnetic bubble films under large in-plane fields, *Phys. Stat. Sol. (a)*, 54, 429-36. [96]·

O'Handley, R.C. (1975), Domain wall kinetics in soft ferromagnetic metallic glasses, *J. appl. Phys.*, 46, 4996-5001. [6, 169-70, 189]

Okabe, Y. (1978), A model to predict the upper cap-switch field of various capping layers, *IEEE Trans. Magnetics*, 14, 602-4. [148]

Olson, C.D., and Pohm, A.V. (1958), Flux reversal in thin films of 82% Ni 18% Fe, *J. appl. Phys.*, 29, 274-82. [194]

Opechowski, W., and Guccione, R. (1965), Magnetic symmetry, in *Magnetism*, Vol. 11A, ed. G.T. Rado and H. Suhl, (Academic Press, New York) 105-65. [68]

Oudet, X. (1974), Saturation magnetic moments at $0\,^{\circ}K$ of $PrY_2Fe_5O_{12}$ and $NdY_2Fe_5O_{12}$ garnets, *Phys. Lett.*, 48A, 252. [31]

Oudet, X. (1979), Valency, ionicity and electronic configuration in rare earths, *J. Physique*, 40, Supplément au no 5, Mai 1979, C5-3957. [31]

Palmer, W., and Willoughby, R.A. (1967), On the velocity of a domain wall in an applied magnetic field, *IBM J. Res. Dev.*, 11, 284-90. [64]

Patterson, R.W., Braginski, A.I., and Humphrey, F.B. (1975), Bubble deformations during translation, *IEEE Trans. Magnetics,* 11, 1094–6. [24]

Patton, C.E. (1968), Linewidth and relaxation processes for the main resonance in the spin wave spectra of Ni-Fe alloy films, *J. appl. Phys.,* 39, 3060–8. [175]

Patton, C.E. (1973), Transit-time limited wall motion in Permalloy films, *Appl. Phys. Lett.,* 22, 317–18. [173]

Patton, C.E., and Humphrey, F.B. (1966), Mobility and loss mechanisms for domain wall motion in thin ferromagnetic films, *J. appl. Phys.,* 37, 4269–74. [169–70, 173, 175, 177, 180]

Patton, C.E., and Humphrey, F.B. (1968), A comparison of the losses for domain-wall motion and ferromagnetic resonance in thin Ni-Fe alloy films, *J. appl. Phys.,* 39, 857–8. [175]

Patton, C.E., McGill, T.C., and Wilts, C.H. (1966), Eddy current limited domain wall motion in thin ferromagnetic films, *J. appl. Phys.* 37, 3594–8. [169, 176–7]

Patton, C.E., Frait, Z., and Wilts, C.H. (1975), Frequency dependence of the parallel and perpendicular ferromagnetic resonance line-width in Permalloy films, 2-36 GHz, *J. appl. Phys.* 46, 5002–3. [176, 196]

Perekalina, T.M., Askochinski, A.A., and Sannikov, D.G. (1961), Resonance of domain walls in cobalt ferrite, *Soviet Phys., JETP,* 13, 303–7. [57]

Perneski, A.J. (1969), Propagation of cylindrical magnetic domains in orthoferrites, *IEEE Trans. Magnetics,* 5, 554–7. [98, 125]

Potter, R.I., Minkiewicz, V.J., Lee, K., and Albert, P.A. (1975), Dynamic properties of magnetic bubbles in amorphous GdCoCu films, *AIP Conf. Proc.,* 29, 76–7. [21, 136]

Prutton, M. (1964), *Thin ferromagnetic films* (Butterworths, London) [189]

Puchalska, I.B., Jouve, H., and Wade, R.H. (1977), Magnetic bubbles, walls and fine structure in ion-implanted garnets, *J. appl. Phys.,* 48, 2069–76. [119, 146]

Rhein, D. (1974), *Grundlagen der digitalen Speichertechnik,* (Akademische Verlag. Geest und Portig K.-G., Leipzig) 53 *et seq.* [185]

Rijnierse, P.J., and De Leeuw, F.H. (1973), Domain wall dynamics in low-loss garnet films, *AIP Conf. Proc.,* 18, 199–212. [82–4, 87]

Rodbell, D.S., and Bean, C.P. (1955), Influence of pulsed magnetic fields on the reversal of magnetisation in square loop metallic tapes, *J. appl. Phys.,* 26, 1318–23. [169–70, 186]

Rosencwaig, A. (1972), The effect of a second magnetic layer on hard bubbles, *Bell Syst. tech. J.,* 51, 1440–4. [21, 117]

Rossing, T.D. (1963), Resonance line width and anisotropy variation in thin films, *J. appl. Phys.,* 34, 995. [175]

Rossol, F.C. (1969), Domain-wall mobility in rare-earth orthoferrites by direct stroboscopic observation of moving domain walls, *J. appl. Phys.,* 40, 1082–3. [16]

Safiullah, A., and Teale, R.W. (1978), Why is the domain wall mobility in bulk yttrium iron garnet so low? *IEEE Trans. Magnetics*, **14**, 900–2. [64]

Sakurai, Y., Kusuda, T., Konishi, S., and Sugatani, S. (1966), Wall motion and rotational magnetisation in thin Permalloy films, *IEEE Trans. Magnetics*, **2**, 570–5. [196]

Schlömann, E. (1971a), Structure of moving domain walls in magnetic materials, *Appl. Phys. Lett.*, **19**, 274–6. [64]

Schlömann, E. (1971b), Structure and energy of moving domain walls, *AIP Conf. Proc.*, **5**, 160–4. [64]

Schlömann, E. (1972), Critical velocity and mass of domain walls in thin magnetic films, *Appl. Phys. Lett.*, **20**, 190–2. [64, 180]

Schlömann, E. (1974), Wave propagation along domain walls in magnetic films, *IEEE Trans. Magnetics*, **10**, 11–17. [73]

Schlömann, E. (1975), High-speed wall motion in bubble films with in-plane anisotropy, *AIP Conf. Proc.*, **29**, 87–8. [99]

Schlömann, E., (1976), Domain walls in bubble films. IV. High-speed wall motion in the presence of an in-plane anisotropy, *J. appl. Phys.*, **47**, 1142–50. [99, 159]

Schryer, N.L., and Walker, L.R. (1974), The motion of 180° domain walls in uniform dc magnetic fields, *J. appl. Phys.*, **45**, 5406–21. [46, 58, 64, 99]

Seitchik, J.A., Doyle, W.D., and Goldberg, G.K. (1971), Simple method of measuring mobility in cylindrical domain materials, *J. appl. Phys.*, **42**, 1272–4. [15]

Shaw, R.W., Moody, J.W., and Sandfort, R.M. (1974), Dynamic properties of high-mobility garnet films in the presence of in-plane magnetic fields, *J. appl. Phys.*, **45**, 2672–7. [92, 95]

Shubnikov, A.V., and Belov, N.V. (1964), *Colored Symmetry* (Pergamon, Oxford) [67]

Shumate, P.W., and Peirce, R.J. (1973), Lifetime characterization of propagated bubble-data streams, *Appl. Phys. Lett.*, **23**, 204–5. [162]

Shvetsov, V.I., and Antip'yev, G.V. (1973), Determination of the direction of spin rotation in a Bloch domain wall, *Physics Metals Metallogr.*, **35**, No. 3, 194–6. [39]

Sixtus, K.J. (1935), Magnetic reversal nuclei. V Propagation of large Barkhausen discontinuities, *Phys. Rev.*, **48**, 425–30. [3]

Sixtus, K.J., and Tonks, L. (1931), Propagation of large Barkhausen discontinuities, *Phys. Rev.*, **37**, 930–58. [3–5]

Sixtus, K.J., and Tonks, L. (1932), Propagation of large Barkhausen discontinuities II, *Phys. Rev.*, **42**, 419–35. [3]

Slonczewski, J.C. (1968), Induced anisotropy in NiFe films, *IEEE Trans. Magnetics*, **4**, 15–9. [197]

Slonczewski, J.C. (1971), Dynamics of magnetic domain walls, *AIP Conf. Proc.* **5**, 170–4. [86, 89, 98]

Slonczewski, J.C. (1972a), Dynamics of magnetic domain walls, *Int. J. Magnetism*, **2**, 85–97. [86, 89, 99]

Slonczewski, J.C. (1972b), Translational mobility of hard ferromagnetic bubbles, *Phys. Rev. Lett.*, **29**, 1679-82. [99, 114, 133, 135]

Slonczewski, J.C. (1973), Theory of domain-wall motion in magnetic films and platelets, *J. appl. Phys.*, **44**, 1759-70. [86, 89, 99, 159]

Slonczewski, J.C. (1974a), Theory of Bloch-line and Bloch-wall motion, *J. appl. Phys.*, **45**, 2705-15. [89, 99, 120, 133-4, 145, 149]

Slonczewski, J.C. (1974b), Properties of Bloch points in bubble domains, *AIP Conf. Proc.*, **24**, 613-14. [99, 134, 136, 148-9]

Slonczewski, J.C. (1979), Force, momentum and topology of a moving magnetic domain, *J. Magnetism Magn. Mater.*, **12**, 108-22. [145]

Slonczewski, J. C., and Malozemoff, A. P. (1978), Physics of domain walls in magnetic garnet films, in *Physics of Magnetic Garnets*, ed. A. Paoletti, Course LXX, International School of Physics, Enrico Fermi (North Holland, Amsterdam) 134-95. [133, 145, 149, 163]

Slonczewski, J. C., Malozemoff, A. P., and Voegeli, O. (1972), Statics and dynamics of bubbles containing Bloch lines, *AIP Conf. Proc.*, **10**, 458-77. [22, 24, 104, 122, 124, 133, 135-6, 145]

Smith, A. B., Kestigian, M., and Bekebrede, W. R. (1973), LPE garnet films without hard bubbles, *AIP Conf. Proc.*, **18**, 167-71. [118, 136]

Smith, D. H., and Thiele, A. A. (1973), Static hard bubble measurements on several epitaxial garnet film compositions, *AIP Conf. Proc.*, **18**, 173-7. [118]

Smith, D. O. (1958), Static and dynamic behaviour of thin permalloy films, *J. appl. Phys.*, **29**, 264-73. [14, 194-5]

Snoek, J. L. (1947), *New Developments in Ferromagnetism* (Elsevier, Amsterdam) [8]

Sokolnikoff, I. S. (1951), *Tensor Analysis* (Wiley, New York) [129]

Sommerfeld, A. (1950), *Mechanics of Deformable Bodies* (Academic Press, New York) [33]

Sommerfeld, A. (1952), *Electrodynamics* (Academic Press, New York) [33, 171]

Soohoo, R. F. (1965), *Magnetic Thin Films* (Harper & Row, New York) [189]

Soohoo, R. F. (1978), Ferromagnetic resonance in ion-implanted garnet bubble films, *J. appl. Phys.*, **49**, 1582-4. [119]

Sparks, M. (1964), *Ferromagnetic Relaxation Theory* (McGraw-Hill, New York) [42]

Spencer, E. G., LeCraw, R. C., and Clogston, A. M. (1959), Low temperature line width maximum in YIG, *Phys. Rev. Lett.*, **3**, 32-3. [51-2]

Speriosu, V., Rosenthal, Y. D., Humphrey, F. B., and Kobayashi, T. (1979), Magnetic bubble dynamics in asymmetric half-disk propagator, *IEEE Trans. Magnetics*, **15**, 875-9. [26]

Stein, K. U. (1969), A 1.5×10^5 bit low-current high-speed planar magnetic film memory with high storage density, *IEEE Trans. Magnetics*, **5**, 912-17. [189]

Stein, K. U., and Feldtkeller, E. (1967), Wall streaming in ferromagnetic thin films, *J. appl. Phys.*, **38**, 4401-8. [205-6]

Stewart, K. H. (1950), Domain wall movement in a single crystal, *Proc. phys. Soc.,* **63A**, 761–5. [165]

Stoner, E. C., and Wohlfarth, E. P. (1948), A mechanism of magnetic hysteresis in heterogeneous alloys, *Phil. Trans. R, Soc.,* **240A**, 599–642. [194]

Suzuki, R., and Sugita, Y. (1975), Absence of dynamic conversion of bubbles in Permalloy-coated garnet films of $(YGdTm)_3(FeGa)_5O_{12}$, *Appl. Phys. Lett.,* **26**, 587–9. [137]

Suzuki, R. and Sugita, Y. (1978), New stable state of bubbles containing Bloch lines, *IEEE Trans. Magnetics,* **14**, 210–12. [114]

Suzuki, R., Takahashi, M., Kobayashi, T. and Sugita, Y. (1975), Planar domains and domain wall structures of bubbles in permalloy-coated garnet layers, *Appl. Phys. Letts.,* **26**, 342–4. [119, 137]

Tabor, W. J., Bobeck, A. H., Vella-Coleiro, G. P., and Rosencwaig, A. (1972a), A new type of cylindrical magnetic domain (Bubble Isomers), *Bell Syst. tech. J.,* **51**, 1427–31. [21, 104]

Tabor, W. J., Bobeck, A. H., Vella-Coleiro, G. P., and Rosencwaig, A. (1972b), A new type of cylindrical magnetic domain (Hard bubble), *AIP Conf. Proc.* **10**, 442–57. [104, 114]

Takahashi, M., Nishida, H., Kobayashi, T. and Sugita, Y. (1973a), Hard bubble free garnet epitaxial films: The garnet permalloy composite structure, *J. Phys. Soc. Japan,* **34**, 1416. [118]

Takahashi, M., Nishida, H., Kobayashi, T. and Sugita, Y. (1973b) Hard-bubble-free garnet-permalloy composite films, *AIP Conf. Proc.,* **18**, 172. [118]

Tarui, Y., Hayashi, Y., Koyanagi, T., Yamamoto, H., Shiraish, M., and Kurosawa, T. (1969), A 40ns 144-Bit n–Channel MOS LSI Memory, *IEEE J. Solid State Circuits,* **4**, 271–9. [189]

Telesnin, R. V., Ilyicheva, E. N., Kanavina, N. G., Stepanova, N. B., and Shishkov, A. G. (1969), Domain-wall motion in thin permalloy films in pulsed magnetic fields, *IEEE Trans. Magnetics,* **5**, 232–6. [173, 180]

Telesnin, R. V., Zimacheva, S. M., and Randoshkin, V. V. (1977), Investigation of the motion of domain walls in iron garnet films, *Soviet Phys., Solid St.,* **19**, 528–9. [158]

Thiele, A. A. (1969), The theory of cylindrical magnetic domains, *Bell. Syst. tech. J.,* **48**, 3287–335. [100]

Thiele, A. A. (1970), Theory of static stability of cylindrical domains in uniaxial platelets, *J. appl. Phys.,* **41**, 1139–45. [100]

Thiele, A. A. (1971), Device implications of the theory of cylindrical magnetic domains, *Bell Syst. tech. J.,* **50**, 725–73. [98, 100, 102, 113]

Thiele, A. A. (1973a), Excitation spectrum of magnetic domain walls, *Phys. Rev. B,* **7**, 391–7. [29, 99]

Thiele, A. A. (1973b), Steady-state motion of magnetic domains, *Phys. Rev. Lett.,* **30**, 230–3. [99, 114, 127–9, 133, 135]

Thiele, A. A. (1974), Applications of the gyrocoupling vector and dissipation dyadic in the dynamics of magnetic domains, *J. appl. Phys.,* **45**, 377–93. [99, 133–4, 143, 145, 149]

Thiele, A. A. (1975), Excitation spectrum of a magnetic domain wall containing Bloch lines, *Phys. Rev. B,* **14**, 3130-65. [99]

Thiele, A. A. (1976), On the momentum of ferromagnetic domains, *J. appl. Phys.,* **47**, 2759-60. [99]

Thiele, A. A., Hagedorn, F. B., and Vella-Coleiro, G. P. (1973), Dynamic spin configuration for hard magnetic bubbles in translational motion, *Phys. Rev. B,* **8**, 241-5. [99, 114]

Thompson, D. A., and Chang, H. (1966), Nanosecond microscopic measurement of magnetic film switching by Kerr magneto-optic apparatus, *Phys. Stat. Sol.,* **17**, 83-90 [15]

Timakova, G. P., Yudina, L. A., Klenin, S. A., Yudin, V. V., and Veter, V. V. (1977), Fourier optics in thin ferromagnetic films, *Physics Metals Metallogr.,* **44**, 63-70. [197]

Tonks, L., and Sixtus, K. J. (1933a), Propagation of large Barkhausen discontinuities III. Effect of circular field with torsion, *Phys. Rev.,* **43**, 70-80. [3, 201]

Tonks, L., and Sixtus, K. J. (1933b), Propagation of large Barkhausen discontinuities IV. Regions of reversed magnetisation, *Phys. Rev.,* **43**, 931-40. [3]

Tsang, C. H., and White, R. L. (1974), Observations of domain wall velocities and mobilities in $YFeO_3$, *AIP Conf. Proc.,* **24**, 749-50. [13, 68]

Tsang, C. H., White, R. L., and White, R. M. (1975), Bloch, Néel and head-to-head domain wall mobilities in $YFeO_3$, *AIP Conf. Proc.,* **29**, 552-3. [12, 68]

Tsang, C. H., White, R. L., and White, R. M. (1978), Domain wall mobilities in $YFeO_3$, *J. appl. Phys.,* **49**, 1838-40. [12, 68]

Uchiyama, S., Shiomi, S., and Fujii, T. (1976), Effect of acoustic wave on domain wall velocity, *AIP Conf. Proc.,* **34**, 154-6. [68]

van der Pol, B. (1920), Discontinuities in the magnetisation, *Proc. Acad. Sci. Amsterdam,* **23**, 637-43 and 980-8. [2]

Vella-Coleiro, G. P., (1972), Domain wall mobility in epitaxial garnet films, *AIP Conf. Proc.,* **10**, 424-41. [21, 42, 136]

Vella-Coleiro, G. P. (1973), Dynamic conversion effects in epitaxial garnet films, *AIP Conf. Proc.,* **18**, 217-21. [161]

Vella-Coleiro, G. P. (1974), Domain wall velocity during magnetic bubble collapse, *AIP Conf. Proc.,* **24**, 595-7. [138]

Vella-Coleiro, G. P. (1975), Overshoot in bubble translation, *AIP Conf. Proc.,* **29**, 64. [25, 137]

Vella-Coleiro, G. P. (1976a), Time-dependent translational velocity of magnetic bubble domains, *Appl. Phys. Lett.,* **28**, 743-5. [25, 138, 162-3]

Vella-Coleiro, G. P. (1976b), Overshoot in the translational motion of magnetic bubble domains, *J. appl. Phys.,* **47**, 3287-90. [25, 137-8]

Vella-Coleiro, G. P. (1976c), Walker-type velocity oscillations of magnetic domain walls, *Appl. Phys. Lett.,* **29**, 445-7. [25, 138, 158]

Vella-Coleiro, G. P. (1976d), Time dependent effects in magnetic domain wall dynamics, *AIP Conf. Proc.,* **34**, 144. [25]

Vella-Coleiro, G. P., and Tabor, W. J. (1972), Measurement of magnetic bubble mobility in epitaxial garnet films, *Appl. Phys. Lett.*, **21**, 7-8. [20-21, 111]

Vella-Coleiro, G. P., Smith, D. H., and Van Uitert, L. G. (1971), Domain wall mobility in some rare-earth iron garnets, *IEEE Trans. Magnetics*, **7**, 745-7. [53]

Vella-Coleiro, G. P., Smith, D. H., and Van Uitert, L. G. (1972a), Resonant motion of domain walls in yttrium gadolinium iron garnet, *J. appl. Phys.*, **43**, 2428-30. [53, 57]

Vella-Coleiro, G. P., Rosencwaig, A., and Tabor, W. J. (1972b), Dynamic properties of hard magnetic bubbles, *Phys. Rev. Lett.*, **29**, 949-52. [99, 111]

Vella-Coleiro, G. P., Smith, D. H., and Van Uitert, L. G. (1972c), Damping of domain wall motion in rare-earth iron garnets, *Appl. Phys. Lett.*, **21**, 36-7. [53]

Vella-Coleiro, G. P., Hagedorn, F. B., Chen, Y. S., and Blank, S. L. (1973), Velocity of domain walls in an epitaxial yttrium-europium garnet film, *Appl. Phys. Lett.*, **22**, 324-5. [20]

Vella-Coleiro, G. P., Hagedorn, F. B., Chen, Y. S., Hewitt, B. S., Blank, S. L., and Zappulla, R. (1974), Magnetic bubble mobility in epitaxial $Sm_{0.4}Y_{2.6}Ga_{1.2}Fe_{3.8}O_{12}$, *J. appl. Phys.*, **45**, 939-42. [20]

Vella-Coleiro, G. P., Hagedorn, F. B. and Blank, S. L. (1975), Dynamic conversion and bubble strip-out, *Appl. Phys. Letts.*, **26**, 67-71. [136-7, 163]

Vella-Coleiro, G. P. Hagedorn, F. B., Blank, S. L., and Luther, L. C. (1979), Coercivity in 1.7 μm bubble garnet films, *J. appl. Phys.*, **50**, 2176-8. [127]

Voegeli, O., and Calhoun, B. A. (1973), Domain formation and associated wall states, *IEEE Trans. Magnetics*, **9**, 617-21. [22-3, 104, 136, 143-5]

Voegeli, O., Calhoun, B. A., Rosier, L. L. and Slonczewski, J. C. (1974), The use of bubble lattices for information storage, *AIP Conf. Proc.*, **24**, 617-9. [146]

de Waard, R. H. (1927), On a theory of the magnetic properties of iron and other metals, *Phil. Mag.*, **4**, 641-67. [3]

Walker, L. R. (1957), Magnetostatic modes in ferromagnetic resonance, *Phys. Rev.*, **105**, 390-9. [49]

Walling, J. C. (1979), Quenching of coercive force in bubble films by domain wall restoring forces, *J. appl. Phys.*, **50**, 2179-81. [126-7]

Walsh, T. J., and Charap, S. H. (1974), Novel bubble drive, *AIP Conf. Proc.*, **24**, 550-1. [205]

Wanas, M. A. (1967), Domain wall motion in yttrium iron garnets, *J. appl. Phys.*, **38**, 1019-21. [11]

Weber, R. (1968), Spin wave resonance, *IEEE Trans. Magnetics*, **4**, 28-31. [175, 197]

Weiss, P. (1907), L'Hypothése du champ moléculaire et la propriété ferromagnetique, *J. Phys. theor. appl.*, **4E**, serie 6, 661-90. [2]

Williams, H. J., Shockley, W., and Kittel, C. (1950), Studies of the propagation velocity of a ferromagnetic domain boundary, *Phys. Rev.*, **80**, 1090-4. [8, 10, 165, 169-70, 172, 180-2, 189]

Wohlleben, D. (1967), Diffraction effects in Lorentz microscopy, *J. appl. Phys.,* **38**, 3341-52. [39]

Wolfe, R., and North, J. C., (1972), Suppression of hard bubbles in magnetic garnet films by ion-implantation, *Bell Syst. tech. J.,* **51**, 1436-40. [21, 117]

Wolfe, R., and North, J. C. (1974), Planar domains in ion-implanted magnetic bubble garnets revealed by Ferrofluid, *Appl. Phys. Lett.,* **25**, 122-4. [146]

Wolfe, R., North, J. C., and Lai, Y. P. (1973), Suppression of hard bubbles in magnetic garnet films by ion-implantation: Dependence on ion species, dose, energy, and annealing, *Appl. Phys. Lett.,* **22**, 683-5. [117]

Yoshimura, H., Hirai, M., Asaoka, T. and Toyoda, H. (1978), A 64-K bit.MOS RAM, *ISSCC Digest Technical Papers,* **XXI**, 148-9. [189]

Zaitzev, A. N. (1976), Possible mechanism of interaction between carriers and domain walls in magnetic semiconductors, *Soviet Phys. Solid St.,* **18**, 72-4. [202]

Zenneck, J. (1902), Uber inductiven magnetischen Widerstand, *Annln Phys.,* **9**, 497-521. [2]

Zhakov, S. V., and Filippov, B. N. (1974), On the theory of electromagnetic losses in monocrystalline ferromagnetic sheets with a domain structure, *Physics Metals Metallogr.* **38**, 14-21. [165]

Zimmer, G. J., Morris, T. M., and Humphrey, F. B. (1974a), Transient bubble and strip domain configurations in magnetic garnet materials, *IEEE Trans. Magnetics,* **10**, 651-4. [24]

Zimmer, G. J., Morris, T. M., Vural, K., and Humphrey, F. B. (1974b), Dynamic diffuse wall in magnetic bubble garnet material, *Appl. Phys. Lett.,* **25**, 750-3. [25]

Zimmer, G. J., Gál, L., and Humphrey, F. B. (1975), Initial rapid domain wall motion in magnetic bubble garnet materials, *AIP Conf. Proc.,* **29**, 85-6. [138]

Zvezdin, A. K., and Matveev, V. M. (1972), Some features of the physical properties of rare earth garnets near the compensation temperature, *Soviet Phys. JETP,* **35**, 140-5. [31]

Subject Index

Acoustic wave 68
Amorphous films 202
Angular momentum 40
Anisotropy constant 29
Anisotropy field 28–30, 57
Automotion 151

Bitter pattern 3, 39
Bloch line 105
   horizontal 105
   vertical 105
Bloch point 148, 153, 155
Bloch wall 7, 38
Bubble collapse 18, 102, 105, 108
Bubble deflection 22, 112, 127, 132, 141, 145
Bubble domain 2, 16, 98
Bubble domain device 26, 69, 102
Bubble domain dynamics 20, 24, 98
Bubble domain materials 69
Bubble state switching 155
Bubble states 21, 104, 146, 149, 151, 153, 155, 164
Bubble translation 20, 111, 134, 139, 151, 154
   high drive 156, 163

Characteristic length 66, 96, 102
Chromium bromide 40
Cluster 142, 145, 152, 163
Clustering 120
Cobalt 27
Coercive force 11, 125, 174
Coercivity 5, 126, 145, 152, 173
Coincident current 190
Coherent rotation 191, 194
Colloid 10, 119
Compensation 21
Computers 13
Conducting bubble domain films 199
Conducting media 7, 64, 165
Conductivity 2, 166, 171
Conductivity tensor 204
Coordinates 61
Creep 146, 190
Cycle time 14

Damping 41
Damping constant 42
Deflection angle 113, 127, 135, 138, 143, 145, 156
Demagnetising effects 177
Demagnetising field 47
Disc files 16
Dispersion 27
Domain 4
   multifingered 114
   stripe 115
Domain drag 202
Domain nucleation and growth 185
Domain theory 34
Domain wall 6, 8, 35, 42, 72, 165, 177, 199, 201, 205
   bowing and bending 64–5, 165
   contraction 64
   inertia 55, 89, 187
   mass 53, 91, 95, 187
   mobility 6, 51, 98, 109, 125, 138, 169, 174, 204
   motion 5, 42, 180
   oscillations 91, 95
   resonance 57
   streaming 205
   structure 7–8, 27, 34, 37, 70, 97, 112, 117, 123, 147, 157, 159, 179
Döring mass 53, 95, 187
Dumb-bell domains 123
Dynamic conversion 161

Eddy current 5, 165, 179, 187
Eddy current damping 7, 168, 185
Electron spin 40
Elliptical domains 122
Evaporated films 18
Exchange constant 33
Exchange field 30
Exchange stiffness 202

Faraday effect 15
Ferrite 8, 57
Ferrite cores 14, 183
   square loop 185